CLOUDS

The Earth series traces the historical significance and cultural history of natural phenomena. Written by experts who are passionate about their subject, titles in the series bring together science, art, literature, mythology, religion and popular culture, exploring and explaining the planet we inhabit in new and exciting ways.

Series editor: Daniel Allen

In the same series

Air Peter Adey
Cave Ralph Crane and Lisa Fletcher
Clouds Richard Hamblyn
Desert Roslynn D. Haynes
Earthquake Andrew Robinson
Fire Stephen J. Pyne
Flood John Withington
Gold Rebecca Zorach
and Michael W. Phillips Jr
Islands Stephen A. Royle
Lightning Derek M. Elsom

Meteorite Maria Golia
Moon Edgar Williams
Mountain Veronica della Dora
Silver Lindsay Shen
South Pole Elizabeth Leane
Storm John Withington
Tsunami Richard Hamblyn
Volcano James Hamilton
Water Veronica Strang
Waterfall Brian J. Hudson

Clouds

Richard Hamblyn

REAKTION BOOKS

Published by Reaktion Books Ltd
Unit 32, Waterside
44–48 Wharf Road
London N1 7UX, UK
www.reaktionbooks.co.uk

First published 2017

Printed and bound in China

A catalogue record for this book is available from the British Library

ISBN 978 1 78023 723 7

CONTENTS

Introduction: 'Airy Nothings'

I look about; and should the guide I choose
Be nothing better than a wandering cloud,
I cannot miss my way.
William Wordsworth, *The Prelude*, Book One

When the playwright Aristophanes set out to satirize the state of fifth-century Athenian philosophy, he cast a chorus of clouds as the source of the 'airy' thinking that was the principal target of his scorn. 'They nourish the brains of the whole tribe of sophists', as the Socrates of his dark comedy *The Clouds* (*c.* 420 BC) declares. 'They are the celestial Clouds, the patron goddesses of the layabout. From them come our intelligence, our dialectic and our reason; also, our speculative genius and all our argumentative talents.'[1] The trouble with sophists and poets 'and other such dirty long-haired weirdies', he complains, is that they were 'always talking about clouds and things':

> That's right, isn't it? 'Deadly lightning, twisted bracelet of the watery Clouds' – and 'Locks of the hundred-headed Typhon' – and 'Conflagrating storms' – and 'Airy nothings' – and 'Crook-talon'd birds, the swimmers of the air' – and – let me think – yes! – 'Showers of moisture from the dewy Clouds'. That's right![2]

But then philosophers have always had their heads in the clouds, their thought processes variously characterized as 'cloudy', 'woolly', 'airy', 'vague', 'nebulous', 'foggy', or lost in clouds of theory. While overly complex thinking serves to 'cloud the issue', wrong-headed thinking leads you into Aristophanes's 'cloud cuckoo land'. But while clouds are ready metaphors for muddle and uncertainty – 'the cloud of unknowing', 'the fog of war'

The wind was 'all bangs, howls, rattles'. A tropical storm.

– they are also agents of the imagination. Leonardo da Vinci, in his *Treatise on Painting*, described clouds and other formless traces as triggers of visual invention, their very formlessness and lack of definable boundaries raising troubling questions about their physical nature, with their mix of solid, liquid and gaseous ingredients confounding the kind of elemental categories that typically prevail on earth. What are clouds, exactly? Objects? Phenomena? Systems? Processes? Do they even count as 'things'? 'Clouds are confusing,' notes the literary scholar Mary Jacobus, 'not so much because they mix elements, or constantly change shape, but because they challenge the phenomenology of the visible.'[3] They confound the senses and trouble the mind; for René Descartes, in his essay, *On Meteors* (1637), clouds constitute the most extreme manifestation of the ungraspable, so if you can philosophize about clouds, he argued, you can philosophize about anything:

> Since one must turn his eyes toward heaven to look at them,
> we think of them as being so high that even poets and

Nicolas Poussin, *Blind Orion Searching for the Rising Sun*, late 1650s, oil on canvas. The huntress Diana (goddess of the moon) is shown standing upon the grey clouds that wreathe Orion's face, literally clouding his vision.

painters see them as the throne of God . . . That makes me hope that if I explain their nature here so that one will no longer have occasion to admire anything about what is seen or descends from above, one will easily believe that it is possible in some manner to find the causes of everything wonderful above the earth.[4]

Clouds, in other words, should be brought down to earth, their varied movements accounted for and thereby rendered unmysterious. But clouds have always been more than merely meteorological, possessing rich cultural and emotional associations that extend far beyond their fleeting atmospheric lives. When the sixth-century mountain hermit and Daoist scholar Tao Hongjing declined his emperor's invitation to serve in the court at Liang, China, he invoked the incommunicability of clouds as a symbol of his distance from the emperor's materialist world:

What is there in the mountains?
Over the ranges there are copious white clouds
But I can only enjoy them by myself
And am unable to offer them as a gift to you.[5]

A circumzenithal arc (also known as a 'cloud smile') is an optical phenomenon caused by the refraction of light through ice crystals present in thin cirriform clouds. Because ice crystals refract sunlight more effectively than water droplets, the colours of a circumzenithal arc are often brighter and more intense than those of a rainbow.

Cloud fantasia:
a ballerina glimpsed
in the clouds, from
Puck magazine, 1911.

For Tao, the clouds represent freedom, purity and immateriality. They are an affirmation of the contemplative life, 'a towering and wonderful foreign land above our heads', in the words of the German philosopher Ernst Bloch, who invoked the imaginative lure of the ultimate wish-landscape above, a fairy-tale realm 'where there are castles, too, taller than they are on earth'.[6]

Clouds have been objects of delight and fascination throughout human history, their fleeting magnificence and endless variability providing food for thought for scientists and daydreamers alike. As this book sets out to show, life without clouds would be physically unendurable – alongside their rain-bearing function, clouds act as a finely tuned planetary thermostat – while it would also be mentally and spiritually bereft, depriving us of a limitless source of creative contemplation. 'Clouds are thoughts without words', as the Canadian poet

Mark Strand observed in a recapitulation of an early Christian conceit that viewed thoughts as clouds centuries before cartoonists invented the now ubiquitous thought bubble.[7] In an Old English dialogue (the prose *Solomon and Saturn*), quoted by Alexandra Harris in her luminous book *Weatherland*, the answer to the question, 'What was Adam made from?' was 'eight pounds of material'. This includes a pound of earth for his flesh, a pound of fire for his blood, a pound of wind for his breath and a pound of cloud (*'wolcnes pund'*, as it reads in the original) from which was fashioned his instability of mind. God, in other words, 'takes a handful of cloud and shapes it into thoughts – except that, being cloud, the thoughts keep changing shape'.[8] Clouds, like thoughts, are thus inherently unreliable but, as the following chapter will argue, they are also an endlessly mutable, magical screen onto which dreams, ideas, myths and legends have always been projected.

Clouds as an aerial projector screen: a promotional illustration from *La Science illustrée* (1894).

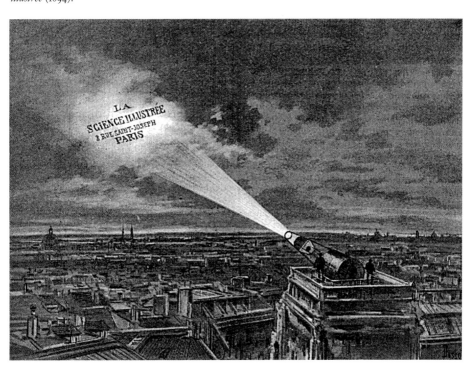

1 Clouds in Myth and Metaphor

'Tell me, Alvis! You're the dwarf who knows everything about our
fates and fortunes: what is the name for the clouds, that hold the
rain, in each and every world?'
    'Men call them Clouds,' the dwarf replied. 'The gods say Chance
of Showers and the Vanir say Wind Kites. The giants name them
Hope of Rain, the elves Weather Might, and in Hel they're known
as Helmets of Secrets.'
Kevin Crossley-Holland, trans., 'The Lay of Alvis'[1]

Some of the world's earliest written documents grapple with the
mysteries of clouds and weather. Ancient Egyptian and Baby-
lonian texts, preserved on papyrus or clay tablets, are filled with
foreboding over rains and droughts. 'When a dark halo surrounds
the moon, the month will bring rain or gather clouds', says a
4,000-year-old Chaldean prophecy; 'When a cloud grows dark
in heaven, a wind will blow', warns another. By the fifth century
BC, weather bulletins were being displayed in the public squares
of cities across the Mediterranean, filled with astro-meteorological
predictions such as 'September 5: Rising of Arcturus: south wind,
rain and thunder', or 'September 12: The weather will likely
change'. Shang-era Chinese scholars were already keeping
detailed weather journals in which sightings of rainbows and
haloes were logged, along with rainfall and prevailing winds.[2] The
harmonious doctrine of yin and yang had been established by
the end of the fourth century BC, with yin associated figuratively
with clouds and rain (as earthbound elements of the female
principle), and yang associated with fire and the heat of the sun
(as celestial elements of the balancing male principle). It is the
warmth of the (yang) sun that nourishes the (yin) clouds through
the mysterious agency of evaporation, while an overabundance
of (yin) rain calls up a compensating bolt of (yang) fire to balance
things out in the teeming atmosphere.
    A few centuries later, Daoist clerics invoked a divine admin-
istration known as the Ministry of Thunder, complete with a
god of lightning, an earl of wind and the master of rain, whose

apprentice, a dissolute minor deity named T'un Y'un ('The Little Boy of the Clouds'), was given the job of shepherding the rain-bearing clouds. It was he who was responsible for their unpredictable movements, and for the sudden unexpected downpours that soaked the people below. The Eight Immortals of Daoist mythology were usually depicted travelling on small flying clouds, although in one Ming-era text from the fifteenth century the Immortals abandon 'their usual mode of celestial locomotion – by taking a seat on a cloud' and demonstrate the scope of their talents by placing objects on the surface of the sea and travelling on them instead.[3] Meanwhile, as the meteorologist Ralph Abercromby noted in an 1887 lecture on clouds in folklore, Daoist mythmakers conjured airborne dragons from the turbulence of passing thunderclouds:

> 'AD 1608, 4th moon. A gyrating dragon was seen over the decorated summit of a pagoda; all around were clouds and fog, the tail only of the dragon was visible; in the space of eating a meal it went away, leaving the marks of its claws on the pagoda.'
>
> These manifestly refer to the long narrow funnel, or tail-shaped cloud, which constitutes the spout of a tornado or whirlwind.[4]

Clouds as a means of divine transportation are also a feature of the Buddhist tradition. A story from the early life of Buddha tells how he summoned a small cloud which obediently ferried him over the Ganges, while the Buddhist cosmology contains a tenfold cloud taxonomy whose stages mirror the tenfold ascent towards enlightenment:

> namely: 1) the great bright clouds of perfection; 2) the great bright clouds of mercy and compassion; 3) the great bright clouds of wisdom; 4) the great bright clouds of Prajna; 5) the great bright clouds of Samadhi; 6) the great bright clouds of Srivatsa; 7) the great bright clouds of blissful virtues; 8) the great bright clouds of meritorious

The Buddha, needing to cross the Ganges, summons a small cloud which dutifully ferries him over. Anon. Chinese illustrator, 1920s.

virtues; 9) the great bright clouds of refuge; and 10) the great bright clouds of praise.[5]

Could this have been the origin of the happy phrase 'to be on cloud nine', meaning, in this case, to be one step away from enlightenment?

As the boundary between the world above and the world below, clouds have always been the haunt of myths and deities. The 'cloud damsels' of Sri Lanka, who emerge from the painted mists of the Sigiriya Rock in a series of well-preserved fifth-century frescoes, are thought to represent the celestial nymphs of Mount Kailash, the revered Tibetan abode of the gods, while in the old Norse pantheon, Odin's wife, Frigg, sits in the Hall of the Mists, spinning the golden threads to be woven by the winds into morning and evening clouds. These high cirriform clouds were her special preserve, untouched by the hands of other gods, such as the uncouth sons of Bor, who flung Ymir the frost giant's brains into the air 'and turned them into every kind of cloud'.[6]

The ancient Greeks were just as preoccupied with the weather's moods and appearances, and their philosophers embarked on enquiries into 'brontology' (the study of thunder), 'ceraunics' (the study of lightning) and 'nephology' (the study of clouds). Thales of Miletus (*c.* 625–545 BC), who is often referred to as Europe's first scientist, was an impressive meteorological theorist; like the ancient Chinese (whose ideas may well have travelled west along the trade routes), he entertained a semi-mystical reverence for water as the sustainer of life on earth. This reverence, combined with the Homeric belief that the earth was held afloat by a vast aqueous bed, led him to picture a universe based entirely on water, a mobile medium that rose and fell between heaven and earth. Thales's aqueous ideas attracted many followers, most notably a fellow Milesian, Anaximander (*c.* 610–547 BC), author of one the world's first scientific treatises, in which he argued that lightning is caused by friction building up inside turbulent clouds. This was an impressive insight, given that these pre-Socratic philosophers had no instruments with which to test or develop their ideas.

Krishna and Radha in the Rain, 18th century, gouache. Krishna, a much-revered Hindu deity, is described in the *Mahabharata* as being the colour of a rain-dark cloud.

One of the celestial 'cloud damsels' of Sri Lanka, depicted emerging from the mist in a well-preserved 5th-century fresco at the Sigiriya complex.

They did, however, have a complex mythology in which clouds were more than just the seats of the gods. Nephele (from *nephos*: 'cloud') was a nymph whom Zeus had shaped from a cloud into the image of his wife, Hera, as a trap to ensnare Ixion, the king of the Lapiths, who had attempted to force himself upon Zeus' queen. The ruse worked, and Ixion raped the cloud-in-disguise, who subsequently gave birth to the race of centaurs (in this version of the story, all centaurs are the offspring of Ixion and Nephele) in a rain shower upon Mount Pelion. Ixion was aptly punished for his crime by a blast from Zeus' thunderbolt. In another story, told by Ovid in his *Metamorphoses*, Jupiter/Zeus reluctantly arms himself with the thunderbolts that will destroy his lover, Semele, as she beholds the blazing sight of his divine form – the vision that she had tricked the god into promising, but which would prove fatal to her mortal sight. Jupiter weeps as he helplessly 'gather[s] the foggy clouds around him', in the words of Ted Hughes's luminous verse translation:

> Now he piled above him the purple
> Topheavy thunderheads
> Churning with tornadoes
> And inescapable bolts of lightning.
> Yet he did what he could to insulate
> And filter
> The nuclear blast
> Of his naked impact . . .[7]

The Norse goddess Frigg spinning cirrus clouds from golden thread in J. C. Dollman's illustration for H. A. Guerber's *Myths of the Norsemen* (1909).

Antonio da Correggio, *Jupiter and Io*, 1531, oil on canvas. In Ovid's tale from the *Metamorphoses*, Jupiter assumed the form of a raincloud in which to hide from his wife, Juno.

But the doomed Semele was already dead, atomized by Jupiter's implacable Olympian weather.

Clouds come into play once more as Jupiter falls in love with Io, daughter of the king of Argos, disguising himself inside a dark, enveloping rain cloud as a means of evading his jealous wife Juno. The scene, from Ovid, was painted in the 1530s by Antonio da Correggio, as part of the 'Loves of Jupiter' series commissioned by the Duke of Mantua, in which Correggio shows the god's face emerging from a rain-heavy cumulus cloud that clasps the helpless mortal in its damp, misty embrace.

Clouds of prediction

As has already been seen, the idea of weather prognosis is as old as civilization, with clouds – as the only reliably visible portions of the atmosphere – featuring in many of the world's weather sayings. 'When the clouds rise in terraces of white, soon will the country of the corn priests be pierced with the arrows of rain', says a Zuni Indian proverb, while the early Hopi people of Arizona carved rain-cloud petroglyphs onto boulders, perhaps as a means of summoning rain in times of drought.[8] A Zuni prayer song recorded in New Mexico in the 1920s pays homage to the *U'wanami* (water spirits) who are called upon to nourish the land with their rain-bearing clouds:

From wherever you abide permanently
You will make your roads come forth,

Your little wind blown clouds,
Your thin wisps of clouds,
Your great masses of clouds
Replete with living waters,
You will send forth to stay with us.[9]

In San Ildefonso Pueblo, New Mexico, the Cloud Dance is performed every other spring. Dancers wear painted wooden headdresses cut in a stepped design to represent layers of clouds. Pueblo pottery, too, invokes the blessings of rain, with vessels decorated with clouds 'in various shapes and forms, windblown clouds, or clouds pouring a lot of water, symbolically painted', testament to a long-lived culture that understands the potency of clouds and rain.[10]

The books of the Bible are filled with hard-won weather observations, such as: 'When ye see a cloud rise out of the west, straightway ye say, There cometh a shower; and so it is' (Luke 12:54); or 'When it is evening, ye say, It will be fair weather: for

A double rainbow over Berlin's Tempelhof Park. The darkened area between the primary and secondary bows is known as Alexander's Dark Band, after the Greek philosopher Alexander of Aphrodisias, who first described the effect in AD 200.

the sky is red. And in the morning, It will be foul weather to day: for the sky is red and lowring' (Matthew 16:2–3), marking the first appearance in print of the familiar weather saying, 'red sky at night . . .'

In the books of the Old Testament, the journeys of the children of Israel were prompted by the appearances of God in the clouds, whether over the Red Sea, over Mount Sinai, or in anger over the tabernacle following the worship of the Golden Calf, when 'all the people saw the cloudy pillar stand at the tabernacle door' (Exodus 33:10). During their 400-year exile on the plains of Egypt, the Israelites would have been unfamiliar with storm clouds – the air above them would have been too hot and too clean to condense locally risen vapour into clouds, while rain-bearing systems travelling north along the trade winds would have exhausted themselves long before reaching the flatlands of the pharaohs: a region with an average rainfall of less than 5 cm per year. The exodus to the semi-arid Sinai brought them into sudden contact with the seasonal uncertainties of rainfall, along with the unfamiliar sight of towering cumulonimbus clouds, the divine 'pillar of cloud' that appeared to them as soon as they had left their deltaic lowlands (Exodus 13:21). The spectacle of clouds came to symbolize the uncertainty and strangeness of exile for the nomadic authors of the books of Moses, and their anxiety can be heard in a volley of questions asked in the books of Enoch and Job: 'Hath the rain a father?'; 'Who can number the clouds in wisdom?'; 'Can any understand the spreadings of the clouds?'

For a farming culture dependent on regular rainfall, these were questions that went to the heart of humanity's post-diluvial contract with God – 'I do set my bow in the cloud, and it shall be for a token of a covenant between me and the earth' (Genesis 9:13) – and for centuries weather prognosis remained a semi-mystical, occult enterprise, evidenced by the ancient Hebrew term *gnoneetz*, 'a cloud-monger: a diviner by looking up to the clouds.'[11] A popular thirteenth-century treatise, *De prognosticatione temporum* (On Weather-forecasting), by the English scholar and churchman Robert Grosseteste, used complex astrological

School of Giotto di Bondone, *St Francis in Ecstasy*, *c.* 1298, part of a fresco cycle in the Upper Church of San Francesco, Assisi. As a vision of Christ appears in the sky, the saint is levitated by a miraculous cloud.

calculations to predict the weather on specific dates, and quoted from a range of ancient sources, such as *Wind and Weather Signs* by the Greek philosopher Theophrastus: 'Fear not as much a cloud from the land as from the ocean in winter; but in the summer a cloud from a darkling coast is a warning.'[12]

Weather prognostics remained a popular publishing staple, with almanacs and vernacular handbooks routinely selling in the tens of thousands. In his popular English-language almanac *A General Prognostication* (first published in 1553), the mathematician Leonard Digges included a range of everyday weather signs collected from farmers and fieldworkers:

If thick clowdes resemblyng flockes, or rather great heapes of woll, be gatherid in many places, they shewe rayne. Also when grosse, thicke, darke clowdes, right ouer the northe part, or somwhat declining to the west, are close with the earth, immediatly folowyth rayne. If they appeare lyke hylles, somedeale from the earth, a good token of weather ouerpassed. Black clowdes, signifie rayne: white clowdes apperyng in winter, at the Horizon, two or three dayes together, prognosticate colde, and snowe.[13]

William Fulke's *A Goodly Gallerye* (1563), a weather guide based on Aristotelian principles, was full of similar prognostications, advising that 'if in the West about the sunne setting, there apeare a black cloude, it wyl raine that nyghte, because that cloude shall wante heat, to disperse it'. Fulke, however, seemed more interested in the outlandish than the everyday, writing at length about miraculous rains of

John Linnell, *Noah: The Eve of the Deluge*, 1848, oil on canvas, illustrating a passage from Milton's *Paradise Lost* (1667): 'Meanwhile the South-wind rose, and, with black wings/Wide-hovering, all the clouds together drove/From under heaven.'

wormes, frogs, fishes, blood, milke flesh, stones, wheat, iron, wool, brick and quicksilver; for histories make mention, that at diverse times, it hath rained such thinges, whose naturall cause for the most parte, we will goe about to expresse,

notwithstandinge accomptinge them amonge suche wonders, as God sendeth to be considered.[14]

To read Fulke is to recognize that Tudor England was a place where the weather was still perceived as a series of wondrous signs, the meanings of which were more metaphysical than geophysical.

In 1670 John Claridge, the self-styled 'Shepherd of Banbury', published a popular collection of rural weather wisdom, *The Shepheard's Legacy*, which was rediscovered in the mid-eighteenth century and reissued as *The Shepherd of Banbury's Rules to Judge of the Changes of the Weather* (1744), going through many subsequent editions. Claridge's earlier book was purported to have been based on decades' worth of first-hand observations made in the open air, in which everything in nature was read 'as a sort of weather-gage: the sun, the moon, the stars, the clouds, the winds, the mists, the trees, the flowers, the herbs, were employed as instruments of real knowledge'.[15] In the 1744 edition, Claridge's words of pastoral wisdom were enhanced with learned annotations supplied by John Campbell, a London-based hack writer whose elucidations sought to add meteorological depth to the shepherd's picturesque pronouncements:

Clouds. *Small and round, like a dappley-grey, with a north wind* – Fair weather for 2 or 3 days.
This is differently expressed by other authors. My Lord *Bacon* tells us, that if clouds appear white, and drive to the N.W. it is a sign of several days fair weather.
Our old English Almanacks have a maxim to this purpose:
If woolly fleeces spread the Heavenly way,
Be sure no rain disturbs the summer day . . .
There is another English proverb worth remembering:
In the decay of the Moon,
A cloudy morning bodes a fair afternoon.[16]

This combination of folk wisdom and scientific scepticism represents a moment of uneasy transition between the medieval weather almanac and the modern meteorological treatise, despite the fact that much of Campbell's added content turned out to have been plagiarized (all his cloud material had been copied verbatim from the Reverend John Pointer's *A Rational Account of the Weather* (1738), leading to an angry exchange of words in the *Gentleman's Magazine*). But the enduring popularity of the Shepherd of Banbury, even into the twentieth century, speaks of a modern readership that remains receptive to its ancestors' time-honoured stores of vernacular weather knowledge.

Clouds of providence

One of the lesser-known stories from the *One Thousand and One Nights* tells of a pious man whom God had provided with a rain cloud that accompanied him wherever he went, so that he could drink and perform his ablutions at will. This continued until the ascetic slackened in his devotions, at which point God removed his cloud and ceased to answer his prayers. The man was filled with grief and longing, distraught at the loss of his cloud companion. But, as he slept one night, a voice spoke to him in his dream saying:

> If you want your cloud to be restored to you, go to such-and-such a king in such-and-such a place and ask him for his blessing. God Almighty will then return it to you, thanks to the blessings attendant on the king's pious prayers.

The ascetic visited the king and appealed for his help in restoring his cloud:

> At first light the king said: 'My God, this servant of Yours begs You to send him back his cloud, and You have power over all things. My God, show him that You accept his petition and restore the cloud to him.' His wife said: 'Amen,' and at that the cloud appeared in the sky. 'Good news,' said the king to me, and as I took my leave of them and went off, the cloud followed me as before.[17]

Clouds have often been seen as tokens of a divine presence – an Arabic saying for someone who is lucky or blessed is 'his sky is always filled with clouds'. In a scene towards the end of W. G. Sebald's *The Rings of Saturn* (1998), the narrator describes a late-afternoon sky in which

> a gap would open up among the billowing clouds, then the rays of sun would reach down to the earth, lighting up patches here and there, and making a fan-shaped pattern as they descended, of the sort that used to appear in religious pictures, symbolising the presence above us of grace and providence.[18]

Sebald's providential sky was filled with crepuscular rays, beams of sunlight that are scattered and made visible by minute particles and compounds in the lower atmosphere, the most common variety of which is known as 'Jacob's Ladder', after the episode in the book of Genesis where Jacob dreams of a ladder connecting earth to heaven, and on which hosts of angels can be seen ascending and descending. In Hawaii these rays are known as 'the ropes of Maui' and in Sri Lanka as 'Buddha's rays', while in Tudor times, according to William Fulke, they were known as 'the descending of the holy ghost, or our Ladies *Assumption*, because these things are painted after suche a sort'.[19] Jacob's Ladder is not the only religious reference in the meteorological lexicon. The optical effect that often accompanies the Brocken spectre (a giant shadow that appears on a screen of low-lying cloud) is known as a 'glory', due to its resemblance to a saint's halo; and the synoptic weather chart was named by Francis Galton in reference to the Synoptic Gospels, which offered multiple perspectives on the miracles of Christ.

Less well known than Jacob's Ladder are the two upward-streaming varieties of crepuscular ray, both of which were described in memorable detail by the Jesuit poet Gerard Manley Hopkins, himself a lifelong sky-watcher. His first encounter with upward-streaming rays was on the stormy afternoon of 30 June 1866. As he noted in his journal, there were 'thunderstorms

The downward-streaming variety of crepuscular ray known as 'Jacob's Ladder'.

all day, great claps and lightning running up and down. When it was bright betweentimes great towering clouds, behind which the sun put out his shaded horns very clearly and a longish way.'[20] The sunset that evening was particularly serene.

Hopkins's encounter with the second, far rarer, variety, known as 'anticrepuscular rays', led to a correspondence with the journal *Nature*, to which he sent a pair of rhapsodic letters on the subject of 'shadow-beams in the east at sunset', in November 1882 and November 1883. This excerpt is from the second letter:

> Yesterday the sky was striped with cirrus cloud like the swaths of a hayfield; only in the east there was a bay or reach of clear blue sky, and in this the shadow-beams appeared, slender, colourless, and radiating every way like a fan wide open. This lasted from 3.30 to about 4.30. Today the sky was cloudless, except for a low bank in the west; in the east was a 'cast' of blue mist, from which sprang alternate broad bands of rose colour and blue, slightly fringed. I was not able to look for them till about 4.30, when the sun was down, and they soon faded. I have not before seen this appearance so

far north, but on the south coast, where I first saw it, I think it might often be witnessed. It is merely an effect of perspective, but a strange and beautiful one.[21]

Anticrepuscular rays occur in exactly the same way as their downward-pointing counterparts, but instead of converging towards the sun, they appear to be scattering away from it, due to a simple trick of perspective that occurs when the viewer's back is turned against the setting (or recently set) sun, and he or she is looking towards a spot on the opposite horizon, known as the antisolar point. The relative rarity of these rays is due partly to the fact that they can only be seen by turning one's back on a setting sun; and, as Hopkins observed, 'Who looks east at sunset?' The cloud shadows that create crepuscular rays are, in reality, near parallel. It is only a trick of perspective that makes them appear to converge from a single point, as can clearly be seen from aerial photographs taken high above the sunlit clouds.

An even rarer and more dramatic trick of cloud-led perspective is the Brocken spectre, a vision in which the observer's own projected shadow appears distorted and enlarged on a screen of low-hanging cloud. It was first observed during the eighteenth century on the Brocken peak in northern Germany, and soon became a sought-after sight among Romantic-era climbers. Coleridge was one of many who ascended the Brocken in pursuit of the apparition, the most unsettling feature of which was that you could only see your own projection: a pair of climbers standing side by side would each only see one figure – their own cloudy alter ego. For Thomas De Quincey, who wrote at length about the Brocken spectre, the troubling vision was 'but a reflex of yourself; and, in uttering your secret feelings to him, you make this phantom the dark symbolic mirror for reflecting to the daylight what else must be hidden'.[22]

One of the best early descriptions of the Brocken spectre appeared in the *Gentleman's Magazine* in May 1749. It was drawn from an account of Charles-Marie de la Condamine's voyages in South America, during which his party ascended Mt Pambamarca, in present-day Ecuador:

Anticrepuscular rays appear to fan out from a point on the horizon immediately opposite the setting sun. The cloud shadows that create such rays are, in reality, near-parallel. It is only a trick of perspective that makes them appear to converge from a single point.

Upon the dissipation of a cloud, in which they had been involved, they perceived the sun rising, which shone very bright; the cloud passed to that side of them, which was farthest from the sun, and each one saw his own shadow projected upon it, and his own only, because the surface of the cloud was irregular; the cloud was so near them that they could distinguish all the parts of the shadow, the arms, the legs and the head; but their astonishment was greatly increased when they perceived the head to be adorned with a glory, or *Aureolus*, formed of 3 or 4 little concentric crowns, of very vivid colours each, with the same variety as the primary rainbow . . . This was a kind of apotheosis of each spectator, and every one enjoyed a sensible pleasure in seeing himself adorned with all these crowns, without perceiving those of his neighbours.[23]

The crowning 'glory', which often accompanies a Brocken spectre, is an optical phenomenon produced by light scattered back (diffracted) towards its source by a cloud of uniformly sized water droplets. It appears as a sequence of rainbow-coloured rings surrounding the observer's magnified shadow and, like the spectre itself, can only be seen by the isolated viewer it surrounds, reinforcing the uncanniness of the experience.

In James Hogg's *Private Memoirs and Confessions of a Justified Sinner* (1824), the protagonist, George Colwan, makes a morning ascent of Arthur's Seat in Edinburgh, where he is amazed to

The Brocken spectre, a gigantic shadow projected onto the surface of low cloud, from Camille Flammarion's *L'Atmosphère* (1888). In reality, each mountaineer would only have been able to see his own, individual spectre, and not those of his companions.

A Brocken spectre
with accompanying
'glory', photographed
on a misty Italian
mountainside.

behold 'this terrestrial glory, spread in its most vivid hues beneath
his feet':

> George did admire this halo of glory, which still grew wider,
> and less defined, as he approached the surface of the cloud
> . . . On that shadowy cloud was the lovely rainbow formed,
> spreading itself on a horizontal plain, and having a slight
> and brilliant shade of all the colours of the heavenly bow,
> but all of them paler and less defined. But this terrestrial
> phenomenon of the early morn cannot be better delineated
> than by the name given of it by the shepherd boys, 'The little
> wee ghost of the rainbow'.[24]

Hogg claimed to have seen the Brocken spectre several times
while working as a shepherd in the 1780s, and his eyewitness
account of the phenomenon, entitled 'Nature's Magic Lantern',
vividly evoked the uncanniness of the encounter with his cloudy
doppelgänger, as well as the opportunity it afforded for a bout
of critical self-examination:

> I turned my cheek to the sun as well as I could, that I might
> see the devil's profile properly defined in the cloud. It was
> capital! His nose was about half a yard long, and his face at
> least three yards; and then he was gaping and laughing so,

that one would have thought he might have swallowed the biggest man in the country.[25]

His account of the spectre in the *Private Memoirs* was less convincing, given that it's not George's own vast shadow that he sees 'delineated in the cloud', but – impossibly – that of his murderous brother, Robert, who is pursuing him across the city. George's spectre is experienced as 'a kind of covenant or epiphany – a revelation of divine, natural and scientific power', as the literary scholar Meiko O'Halloran notes in a recent book on Hogg, for whom the Brocken spectre proved more alluring as a literary trope than as a natural phenomenon, an oneiric vision to be projected onto the cinema screen of the clouds.[26]

Clouds as impediments

On a chilly October evening in 1743 Benjamin Franklin stood in his Philadelphia garden, preparing to witness a long-awaited lunar eclipse. Much to his disappointment, a thick bank of storm clouds moved in from the west, completely obscuring the night sky. A few days later, he learned that the eclipse had been clearly visible in neighbouring New England, where the storm clouds had not arrived until after the event. Franklin's disappointment was soon forgotten as he realized the implications of this celestially timed long-range weather pattern, and his failure to see the eclipse led instead to his pioneering theory of the cyclonic movement of storms.

Clouds are often cast as the enemies of astronomy. In Walt Whitman's poem 'On the Beach at Night' (1871), a young girl cries when clouds obscure her view of the planet Jupiter, fearing that the distant object has been devoured by the 'ravening clouds, the burial clouds, in black masses spreading'. Her father reassures her that:

The ravening clouds shall not long be victorious,
They shall not long possess the sky, they devour the stars
only in apparition,

> Jupiter shall emerge, be patient, watch again another
> night, the Pleiades shall emerge . . .[27]

swimming back into glorious vision as though through a parting
curtain. The image recalls Caliban's dream from *The Tempest*, in
which

> The clouds methought would open and show riches
> Ready to drop upon me, that when I waked
> I cried to dream again. (*The Tempest*, iii.2)

The meteorologist James Glaisher credited his time at the
Cambridge Observatory for his own emerging interest in clouds.
'Often between astronomical observations,' he recalled, 'I watched
with great interest the forms of the clouds, and often, when a
barrier of cloud had suddenly concealed the stars from view, I
wished to know the cause of their rapid formation.'[28] He soon
abandoned astronomy altogether and devoted himself to the
study of weather instead, going on to encounter clouds at close
range during his record-breaking balloon ascents of the 1860s.
In both Franklin's and Glaisher's cases, astronomy's loss was
meteorology's gain, a thought that would have offered little
comfort to the hapless French astronomer Guillaume Le Gentil,
who, having failed to observe the 1761 transit of Venus in India
due to hostilities between Britain and France, spent eight long
years there in preparation for the next one. The skies of Pondi-
cherry, he noted in early June 1769, had remained cloud-free for
weeks, allowing him to entertain local officials with views of
Jupiter's moons through his 4.5 m telescope. But the morning of
the second transit (4 June 1769) grew suddenly overcast, and all
that day Le Gentil was fated to see nothing but clouds, a mis-
fortune that apparently drove him to the brink of insanity. 'This
is the fate which often awaits astronomers', as he recalled in his
travel memoir, *Voyage dans les mers de l'Inde* (1781):

> I had gone more than ten thousand leagues; it seemed that I
> had crossed such a great expanse of seas, exiling myself from

my native land, only to be the spectator of a fatal cloud which came to place itself before the sun at the precise moment of my observation, to carry off from me the fruits of my pains and of my fatigues . . .[29]

Was there ever a more unfortunate fieldworker than Guillaume Le Gentil, the luckless victim of a 'fatal cloud'?

A final (and happier) example: if heavy clouds had not cleared above the island of Principe during the afternoon of 29 May 1919, the course of twentieth-century science might have taken a different turn. Four years earlier, Albert Einstein had published his general theory of relativity, in which he posited that light from distant stars is 'bent' by the sun's gravitational field. The only realistic way to test this claim was during the total solar eclipse of 1919, when the sun was due to cross the Hyades star cluster, the bright lights of which would offer plenty of chances to measure a solar deflection. An expeditionary team led by the astronomer Arthur Eddington set off to photograph the event from two locations, one in Brazil, the other on the remote island of Principe, in the Gulf of Guinea. On the morning of the eclipse, thick clouds over Principe threatened to scupper yet another scientific expedition, but the sky miraculously cleared just in time. Eddington took dozens of photographs during the six and a half minutes of totality, analysis of which demonstrated that light from the distant stars was indeed shifted by the sun just as Einstein had predicted. Later that year, the results were made public at a meeting of the Royal Society, and by the following morning, Einstein was famous. The *Times* of London ran a banner headline reading: 'Revolution in Science – New Theory of the Universe – Newtonian Ideas Overthrown', while the *New York Times* went with 'Light All Askew in the Heavens: Einstein's Theory Triumphs'. Yet if the clouds above Principe had failed to dissipate that afternoon, just as the eclipse began, Einstein's theory of relativity might never have been proved in his lifetime, leaving $E=mc^2$ to languish as just another long-forgotten equation.

Clouds from above

To rise above the clouds is to enact a powerful mythic reversal, as though turning the earthly elements upside down. Before hot air balloons tore through the clouds in the late eighteenth century, few in Europe had experienced cloudland from above. One of those who did was the diarist John Evelyn, whose rapturous encounter with a bank of Alpine stratus in the summer of 1644 was something he remembered for the rest of his life:

> as we ascended, we enter'd a very thick, solid, and darke body of Clowds, which look'd like rocks at a little distance, which dured us for neere a mile going up; they were dry misty Vapours, hanging undissolved for a vast thicknesse, & altogether both obscuring the Sunn and Earth, so as we seemed to be rather in the Sea than the Clowdes, till we having pierc'd quite through, came into a most serene heaven, as if we had been above all human Conversation, the Mountaine appearing more like a greate Iland, than joyned to any other hills; for we could perceive nothing but a Sea of thick Clowds rowling under our feete like huge Waves . . . This was one of the most pleasant, new & altogether surprizing objects that in my life I had ever beheld.[30]

Passing through the cloud layer meant leaving the human realm altogether, and entering instead 'a most serene heaven' in which earthly considerations had no place. The same godlike perspective was employed by William Wordsworth towards the end of his long autobiographical sequence, *The Prelude* (1850), in which he recalled an ascent of Mt Snowdon that ended with a transcendent view over the cloudscape below:

> . . . at my feet
> Rested a silent sea of hoary mist.
> A hundred hills their dusky backs upheaved
> All over this still ocean; and beyond,

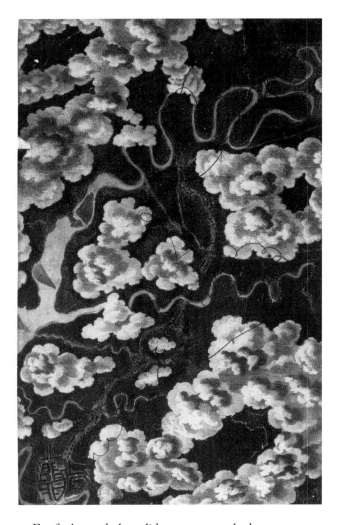

'A Balloon Prospect from Above the Clouds', from Thomas Baldwin's *Airopaidia* (1786).

Far, far beyond, the solid vapours stretched,
In headlands, tongues, and promontory shapes,
Into the main Atlantic . . . [31]

It is a moment of powerful elemental reversal, in which the layer of 'solid' clouds surmounts both land and sea, transfigured into an aerial ocean that drowns the hills and valleys. This kind of rhapsodic encounter with the 'cloud-floor' had already become a staple of early ballooning literature, such as Thomas Baldwin's

Airopaidia (1786), which featured awed descriptions of seas of cloud stretching away beneath the rising basket:

> The lowest bed of Vapour that first put on the appearance of Cloud was of a pure white, in detached Fleeces, increasing as they rose. They presently coalesced, and were aggrandized into a Sea of Cotton, but more white, and dazzling: tufted here and there by the light play of air, and gentle breezes in every direction: but where undisturbed the whole became an extended firmament or white floor of thin cloud . . . thro' this white floor uprose in splendid majesty and awful grandeur, at great and unequal distances, a vast assemblage of *Thunder-Clouds*: each congeries consisting of whole acres in the densest form.[32]

Ballooning might have started out as aerial adventuring, but it soon became an important meteorological tool, with all manner of scientific instruments taken up into the clouds, as well as, on a cross-Channel flight in 1785, the world's first airmail delivery. In the 1860s the meteorologist James Glaisher persuaded the British Association for the Advancement of Science to sponsor a major atmospheric research programme into 'Hygrometric and other Conditions of the Upper Air', the published accounts of which, *Travels in the Air* (1871), transfixed Victorian readers with tales of euphoric ascents made high above the realm of the recently classified clouds:

> A most remarkable view presented itself: the sky was of a fine deep blue, dotted with cirri. The earth and its fields, where visible, appeared very beautiful indeed – here, hidden by vast cumuli and plains or seas of cumulo-strata, causing the country beneath to be shaded for many hundreds of square miles . . . Due north, a beautiful cloud [which] had followed us on our way, still reigned in splendour, and might from its grandeur have been called the monarch of clouds.[33]

During the most celebrated of his flights, on 5 September 1862, Glaisher and his co-pilot, Henry Coxwell, ascended to a height of just over 11 km (37,000 ft) making them the first humans to enter the thin air of the stratosphere. Both nearly died in the attempt, however, Glaisher having passed out from oxygen starvation at a height of nearly 9 km, leaving Coxwell, whose hands had frozen in the sub-zero temperatures, to clamber onto the basket and jam open a broken valve with his teeth. As Richard Holmes observes, the advent of ballooning created 'a new kind of sublime', in which the fear of falling upwards was just as profound as the fear of falling downwards.[34] It also emptied the skies of angels, in the words of the cloud-haunted protagonist of Stéphane Audeguy's novel *The Theory of Clouds* (2005), 'filling it instead with men in balloons or planes'.[35] Audeguy tells the time-travelling story of a Japanese collector of cloud literature and the archivist who is employed to catalogue his vast library in Paris. As the archivist works, the collector describes the history of cloud understanding, from ancient myths and legends to the modern history of classification and forecasting, observing that, since the mid-twentieth century, mass aviation has allowed everyone to experience the clear blue intensity of the sky above the clouds.

The atmosphere is no longer uncharted territory. 'We are the first generation to see the clouds from both sides,' observes

A 1940s promotional poster for Air France: 'Towards New Skies'.

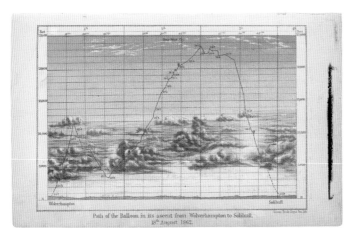

Chart from James Glaisher's *Travels in the Air* (1871), showing both distance travelled and altitude gained during one of Glaisher's balloon ascents.

AIR FRANCE

TOWARDS NEW SKIES

Clouds from above: the Woolworth tower enveloped in a layer of cloud, New York City, 1928.

This view from an aeroplane window reveals two distinct layers of cloud: below, a heavy blanket of stratocumulus spreads as far as the eye can see, while 3 or 4 km (10,000–12,000 ft) above, an icy layer of cirrostratus refracts the incoming sunlight into delicate purples and reds.

the narrator of Saul Bellow's novel *Henderson the Rain King* (1959): 'What a privilege! First people dreamed upward. Now they dream both upward and downward. This is bound to change something.'[36] Certainly for Joni Mitchell, whose breakout song 'Both Sides, Now' (1969) was inspired by Bellow's novel, it changed everything. As she later recalled in an interview with the musician and broadcaster Malka Marom:

But I was reading this on a plane – page twenty-eight,
I think it is. He's up in a plane, looking down on clouds,
and I go, 'Huh! I'm on a plane looking down at clouds.'
I put down the book, I look out the window, and I start
to write.[37]

Mitchell's aerial writing was prompted by the technological miracle of reading above the clouds, a vision of godlike serenity that is memorably reversed in the untitled final poem from Derek Walcott's valedictory collection *White Egrets* (2010), in

The Golden Gate
Bridge emerging from
a canopy of cloud.

which the act of reading is likened to glimpsing the world through a break in the cloud-page: 'This page is a cloud', he writes, between the fraying edges of which a fantastical landscape emerges, headlands, mountains, grooved seas – but the more that is read, the more the mists disperse into clarity, until, inevitably,

> a cloud slowly covers the page and it goes
> white again and the book comes to a close.[38]

Clouds, in this reading, are no longer a mutable, magical screen for our thoughts and ideas, but a final curtain that inexorably descends, enveloping the realm of the imagination in its blank, elemental whiteness.

2 The Natural History of Clouds

Sally: 'How high are the clouds, Linus?'
Linus: 'Oh, they're at different heights. Some of them are "far-away high" and some of them are "right-up-there high".'
Charlie Brown: 'What sort of explanation is that?'
Linus: 'Sometimes it's best to keep these things in the language of the layman.'
Charles M. Schulz, *You Can Do It, Charlie Brown* (1963)[1]

Compared to other branches of natural history, meteorology was a late developer, being established as a subject of systematic inquiry only in the nineteenth century. This was partly due to the practical difficulty of finding and collecting portable specimens. As the British meteorologist William Clement Ley pointed out in a lecture delivered in 1879, 'I cannot bring a cirrus or a cumulus cloud into this room, and then proceed to examine or point out its peculiarities.'[2] Neither can one snap off a piece of rainbow or collect a sample of wind for convenient study indoors. As the study of complex, short-lived phenomena taking place some distance from the observer, meteorology can never hope to be an exact science. It is, at best, a search for narrative order among the fleeting caprices of the atmosphere, the fathomless aerial ocean that the balloonist James Glaisher aptly described as 'the great laboratory of changes'.[3] This chapter will begin by examining some of the earliest systematic attempts to understand the nature of clouds and weather, before moving on to outline the current state of nephological knowledge, and ending with a narrative tour of the ten main types of cloud.

Aristotle's 'exhalations'

Aristotle was the classical world's most influential philosopher-scientist, whose prolific output included the *Meteorologica* (*c.* 340 BC), the first comprehensive treatise on the subject of weather. The treatise begins by dividing the universe into two major spheres: the celestial region, above the moon, and the terrestrial region, below the moon. While the former was the realm of astronomy (the study of the fixed stars), the latter was the realm of meteors – 'any bodies in the air or sky that are of a flux and transitory nature', in the words of Samuel Johnson's *Dictionary*, hence 'meteorology'. Clouds, under this definition, were classified as meteors, as were winds, storms and, curiously, earthquakes.

Aristotle's terrestrial region was composed of four basic elements – earth, air, fire and water – arranged in concentric layers around the earth. These elements, however, were in a constant state of flux, energetically merging and separating, their frequent collisions responsible for the formation of atmospheric 'vapours' and 'exhalations'.[4] The sun's heat, for example, mixed with cold and moist water to form a new airy substance from which clouds and rain formed, while thunder and lightning were caused by dry exhalations trapped inside clouds as they condensed and collided in the atmosphere. The varying sounds of thunder were produced by the varying shapes of clouds.

One of the clearest expressions of Aristotle's ideas which deal with the formation of the universe can be found in the opening verses of Ovid's first-century AD poem *Metamorphoses*:

> Over all these regions hangs the air, as much heavier than
> the fiery aether as it is lighter than earth or water. To the air
> the god assigned mists and clouds, and thunder that was
> destined to cause human hearts to tremble: here, too, he
> placed the thunderbolts, and winds that strike out lightnings
> from the clouds.[5]

Aristotelianism dominated Western science and philosophy for the next two millennia, with many Roman authors setting out to preserve the Greek spirit of inquiry through the compilation of scientific anthologies, such as Seneca's ten-book *Naturales quaestiones*, which he began in around AD 62. For Seneca, clouds were the very engines of weather, being directly responsible for thunder, lightning, rain and rainbows, as well as indirectly in control of the overall shape of the atmosphere: 'Clouds – which are closely associated with atmosphere, into which atmosphere congeals and from which it is dissolved – sometimes gather, sometimes disperse, and sometimes remain motionless.'[6]

It was the Roman poet Lucretius, in his *De rerum natura* (On the Nature of Things, *c.* 50 BC), who first proposed a non-Aristotelian explanation of cloud formation, in which a sudden coalescence in the upper atmosphere sees atoms of rough material band together to form larger, rain-bearing structures:

> Now clouds combine, and spread o'er all the sky,
> When little rugged parts ascend on high,
> Which may be twined, though by a feeble tie;
> These make small clouds, which, driven on by wind,
> To other like and little clouds are joined,
> And these increase by more: at last they form
> Thick, heavy clouds; and thence proceeds a storm.[7]

This passage, taken from Thomas Creech's luminous verse translation of 1682, did much to influence the eighteenth-century revival of Lucretius' atomist model of cloud formation, which arose just as Aristotle's long-held influence finally began to wane.

The rise of nephology

The growth of atmospheric science during the eighteenth and nineteenth centuries saw increased levels of attention paid to the question of cloud formation. Many competing hypotheses were advanced, including the 'menstruum' theory, named after the alchemical term for a solvent, which held that acids in the air were responsible for corroding water into cloud-shapes and keeping them suspended, just as they 'clouded' the surfaces of metals. Another explanation claimed that particles of fire detach from sunbeams and adhere to water droplets, creating lighter-than-air molecules that rise in the air to join together as clouds. Rain is then produced by the separation of the fiery particles, releasing water from the cloud, which then falls under the influence of gravity. In China it was widely believed that 'clouds come out from mountains', but as the Taiwanese meteorologist Pao K. Wang discovered when he set off into the foothills in search of the legendary cloud source, there was 'no "central storage house" of clouds in any mountain in Taiwan'.[8]

The most widespread theory of all was the vesicular, or 'bubble', theory, which maintained that particles of water, through the action of the sun, form hollow spherules filled with a highly rarefied form of air, which, growing lighter than atmospheric air, rise like balloons to form clouds. Rain then falls when these air bubbles 'burst'. It was a plausible hypothesis that appeared to answer the most difficult question of all: if clouds are formed from water – which is a hundred times denser than air – how do they remain suspended in the sky? Such was the influence of the vesicular theory that even Horace-Bénédict de Saussure, the pioneer of Alpine science, claimed to have encountered cloud bubbles during a mountain expedition in the 1780s, the drops floating 'slowly before him, having greater diameters than peas, and whose coating seemed inconceivably thin'.[9]

But none of these explanations endured for long. As Oliver Goldsmith observed in his *History of the Earth, and Animated Nature* (1774), 'Every cloud that moves, and every shower that falls, serves to mortify the philosopher's pride, and to show him hidden qualities in air and water that he finds difficult to explain.'[10] The first true breakthrough in modern meteorology occurred in December 1802, when a young amateur meteorologist named Luke Howard delivered an evening lecture to a London science club, in which he proposed a radical new classification and nomenclature of clouds. The lecture caused an immediate sensation, and within a few years of its publication in 1803, had succeeded in revolutionizing our understanding of the skies.

Howard, who was a pharmacist by profession but a meteorologist by inclination, had been obsessed by clouds and weather since childhood. As a schoolboy he had spent hours staring out of his classroom windows, gazing at the passing clouds. Like everyone else at the time, he had no idea how clouds formed, or how they stayed aloft, but, as he later recalled, his 'observing eye had begun to be attracted by the varying beauties of the cloud-streaked sky'.[11] By his own admission, Howard paid little attention to his lessons, but fortunately for the future of meteorology he managed to pick up a good working knowledge of Latin, which was still the preferred language of communication throughout the scientific world. By the time he gave his celebrated lecture in 1802, Howard had settled in Plaistow, then a village just outside of London, where his daily walks to work had apparently rekindled his earlier interest in the sky: 'Being often *sub dio* in passing to and fro,' he wrote, 'I was brought into a more minute and constant observation of the sky and clouds, resulting in my work entitled "Essay on the Modifications of Clouds, etc.", presented first to the Askesian Society.'[12]

His daily constitutionals may well have played their part, but Howard's ideas had also been shaped by the recent work of the polymath John Dalton, whose *Meteorological Essays* (1793) argued that clouds, far from being Shakespeare's 'airy nothings', obeyed the same laws of physics as everything else in nature. It was Dalton who intuited that once water had condensed from vapour into liquid droplets, it would be acted upon by the same forces that it would meet at ground level. Clouds, in other words, do not 'float' as they appear to do, but instead fall slowly under the influence of gravity. Some of them stay aloft due to upward convection from the sun-heated ground, being formed from tiny droplets with a high degree of air resistance, but most are in a state of slow, balletic descent. This fact explained much about cloud behaviour, particularly the way that high clouds often descend and spread into layers.

Howard's classification was based on the simple but penetrating insight that clouds have many individual shapes but few basic forms; in fact, all clouds belong to one of three principal types, to which Howard gave the names: *cirrus* (Latin for 'fibre' or 'hair'), *cumulus* ('heap' or 'pile') and *stratus* ('layer' or 'sheet'). His schoolroom Latin had at last proved its value.

But that was not the revolutionary part. As Howard pointed out in his talk, clouds are constantly changing, merging, rising, falling and spreading throughout the atmosphere, rarely maintaining the same shapes for more than a few minutes at a time.

a
a
b
c

Lewis. fecit

Engraving showing the three main families of cloud – cirrus, cumulus and stratus – from Luke Howard's 1803 essay 'On the Modifications of Clouds'.

.

Any successful classification would need to accommodate this essential instability. So, in addition to the three main cloud types (cirrus, cumulus and stratus) he introduced a series of intermediate and compound types as a way of marking the regular transitions that occur among the clouds, with their names reflecting their structural transformations. So high, wispy cirrus clouds that had descended and spread into layers were named *cirrostratus*, while groups of fluffy cumulus clouds that had joined up and spread were named *stratocumulus*. A change of form or altitude could thereby be tracked by a change in word order, with the prefix *cirro* denoting a high cloud, *cumulo* a low heaped cloud and *alto* denoting a family of medium clouds. Howard identified a total of seven cloud types in his original lecture, ending with *nimbus*, the rain cloud, which he described as a rainy combination of the three principal types. The system was simple, and once learned, he claimed, clouds would be 'as distinguishable from each other as a tree from a hill or the latter from a lake'.[13]

What was so ingenious about this apparently simple system was that it allowed for the changeability of clouds, while providing a kind of narrative framework, a means of keeping track of clouds as they changed over time. Clouds, in this reading, are not static objects, they are connected stages in a dynamic process, so to observe how a rising bank of stratus cloud 'ascends and evaporates, or passes off with the appearance of a nascent cumulus' is to recognize every cloud's potential for transformation.[14]

It was this elegantly fluid solution to the problem of naming transitional forms in nature that attracted the attention of the great German polymath Johann Wolfgang von Goethe, who came across Howard's classification in an 1815 translation. The timing was fortuitous, as Goethe had recently become preoccupied·with a new area of study, which he termed morphology ('the science of forms'). Goethe's morphology was based on the interconnectedness of all natural patterns and formations, in which objects such as clouds were viewed as manifestations of the same shaping forces responsible for every other dynamic form in nature, such as whirlpools, snowflakes or falling leaves.

There was never a moment in nature, he claimed, when nothing could be said to be happening. Goethe had long been fascinated by clouds and weather – the first entry in his Italian travel diary, for 3 September 1786, saw him nod farewell to the brooding German sky, whose 'upper clouds were like streaked wool, the lower heavy' – and he considered the new cloud classification to be morphology's 'missing thread'.[15] In his rhapsodic essay *Wolken-gestalt nach Howard* (Cloud-shapes according to Howard), he hailed it as a moment of pure, unmediated insight into the 'wholeness' of nature: 'How much the Classification of the clouds by Howard has pleased me,' he wrote, 'how much the disproving of the shapeless, the systematic succession of forms of the unlimited, could not but be desired by me, follows from my whole practice in science and art.'[16]

But the essay was only the start. Goethe's admiration and sense of indebtedness to Howard's classification led to an extraordinary verse homage, *Howards Ehrengedächtnis* (In Honour of

A watercolour sketch by Luke Howard, *c.* 1803–11, showing a layer of cirrocumulus below a patch of high cirrus cloud.

Luke Howard, *Cloud Study*, c. 1803. Howard annotated this watercolour sketch of a small cumulonimbus cloud with the words 'rain hitting the ground, anvil is spread out'.

Howard), in which Goethe set out to explore the mood as well as the mechanics of the three main families of cloud:

Stratus

When o'er the silent bosom of the sea
The cold mist hangs like a stretch'd canopy;
And the moon, mingling there her shadowy beams,
A spirit, fashioning other spirits seems;
We feel, in moments pure and bright as this,
The joy of innocence, the thrill of bliss.
Then towering up in the darkening mountain's side,
And spreading as it rolls its curtains wide,
It mantles round the mid-way height, and there
It sinks in water-drops, or soars in air.

Cumulus

Still soaring, as if some celestial call
Impell'd it to yon heaven's sublimest hall;
High as the clouds, in pomp and power arrayed,
Enshrined in strength, in majesty displayed;
All the soul's secret thoughts it seems to move,
Beneath it trembles, while it frowns above.

Cirrus

And higher, higher yet the vapours roll:
Triumph is the noblest impulse of the soul!
Then like a lamb whose silvery robes are shed,

The fleecy piles dissolved in dew drops spread;
Or gently waft to the realms of rest,
Find a sweet welcome in the Father's breast.[17]

The stanzas, which were first published in German in 1817, stressed the structural connections between the cloud types, just as Howard had described them in his classification, but they also offered acutely observed descriptions of their varying personalities. For Goethe, one of the attractions of the new classification had been that it celebrated aerial nature without attempting to contain it, and while accounting for the material forces of cloud formation, it also allowed the immaterial forces of poetic and emotional response to be heard, thereby transfiguring mankind's relationship with the sky. Goethe sought to express this idea in a final stanza, addressed to Howard, written in 1822 as an explanatory coda to the earlier verses, which had recently appeared in an English translation:

But Howard gives us with his clearer mind
The gain of lessons new to all mankind;
That which no hand can reach, no hand can clasp,
He first has gain'd, first held with mental grasp.
Defin'd the doubtful, fix'd its limit-line,
And named it fitly. – Be the honour thine!
As clouds ascend, are folded, scatter, fall,
Let the world think of thee who taught it all.[18]

The poem led to a brief correspondence between Goethe and Howard, who was disarmed by the attentions of the German colossus. Goethe asked Howard to send an account of his life, 'that I might comprehend how such an intellect had been trained'. The Englishman's modest reply, in which he described the limitations of his schooling – 'my pretensions as a man of science are consequently but slender' – delighted the older man, for whom Howard's character supplied 'no finer example of the sort of mind to which nature delights to reveal herself'.[19] A verse from Goethe's diary confirms the lasting impact that the friendship had on him:

Luke Howard, *Cloud Study*, c. 1808–11, watercolour sketch. This closely observed sketch shows how parallel bands of cirrus fibratus clouds appear to radiate out from the horizon.

Disciple of Howard, strangely
You look around and above you every morning
To see whether the mist falls or rises
And what clouds are showing.[20]

Goethe promoted Howard's classification throughout the German-speaking world, recommending it particularly to artists of his acquaintance. As a fifteen-year-old art student, Friedrich Preller was given a copy of Howard's essay by Goethe, who instructed him to 'read that and then observe the various cloud formations and bring me clear drawings of them'.[21] Preller willingly complied with Goethe's request, and went on to feature cloud-filled skies in much of his later work.

The Dresden-based painters Carl Gustav Carus and Johan Christian Dahl received similar instructions from Goethe, with Carus – who was a leading physician and naturalist as well as a painter – responding especially favourably to *Howards Ehrengedächtnis*. As soon as he read it, he realized that the question of how to reconcile scientific analysis with creative freedom had been answered. In a letter addressed to his son, Ernst, Carus gave an admiring account of Goethe's poem:

The poem on clouds could never have been written without long and hard prior study of the atmosphere; the poet had to observe, judge, distinguish, until he attained not only knowledge of the formation of clouds, as known from the evidence of the senses, but the insight that is the fruit of scientific investigation alone. After all this, the mind's eye brought into focus all the separate rays given off by the

Carl Gustav Carus, *Goethe Monument*, 1832, oil on canvas. An idealized image that relocates Goethe's Weimar tomb to a mist-shrouded mountain setting. Goethe had recommended the study of clouds to Carus, who was a physician and naturalist as well as a noted landscape painter.

Friedrich Preller the Elder, *Storm on the Coast*, 1856, oil on canvas. While a student at Dresden, Preller was encouraged to paint clouds by Goethe, who gave him a translation of Howard's essay.

phenomenon and reflected the essence of the whole in an artistic apotheosis.[22]

His own immersion in the new cloud names can be gauged by a journal entry from October 1823, in which he records how 'one evening, the vapors coalesced in the higher regions of the sky into long strands of cirrostratus', while 'toward the west, horizontal strands of flocculent cloud, lying one behind the other like waves, spanned the vault of the sky'.[23] But not everyone in

Goethe's circle welcomed the new meteorology. Caspar David Friedrich, who was Carus's art tutor at Dresden, objected to what he saw as the scientific attempt 'to force the free and airy clouds into a rigid order and classification', arguing that the obscurity and imprecision of the clouds were valuable attributes in themselves.[24] For Friedrich, whose *Wanderer above the Sea of Fog* (1818) became an icon of German Romanticism, clouds were banners of creative and contemplative freedom as well as portals to the divine realm, and should be left to their own devices.

The cloud-by-cloud structure of Goethe's poem was not the only example of the form. A near-contemporary work by Percy Shelley, 'The Cloud' (1820), offered a vivid poetic primer of Howard's classification, while delivering an atmospheric meditation on human creativity. The seven modifications appear sequentially as a fluid, changeable organism that addresses the reader in an appealingly mocking first-person voice:

I bring fresh showers for the thirsting flowers,
    From the seas and the streams;
I bear light shade for the leaves when laid
    In their noonday dreams.
From my wings are shaken the dews that waken
    The sweet buds every one,
When rocked to rest on their mother's breast,
    As she dances about the sun.
I wield the flail of the lashing hail,
    And whiten the green plains under,
And then again I dissolve it in rain,
    And laugh as I pass in thunder.[25]

This first stanza invokes cumulus, stratus and cumulonimbus clouds, while the poem goes on to characterize cirrus (the moon 'glides glimmering o'er my fleece-like floor'), cirrocumulus ('when I widen the vent in my wind-built tent') and cirrostratus clouds ('I bind the sun's throne with a burning zone, and the moon's with a girdle of pearl'). All the modifications are presented by Shelley as aspects of a single, protean cloud personality, in direct and knowing reference to Howard's contention that clouds unite, pass into one another and disperse in recognizable stages. Shelley's perpetual, mutable cloud, with its intimation of immortality – 'I change, but I cannot die' – is a powerful personification of the entire atmospheric system:

> I am the daughter of Earth and Water,
>   And the nursling of the Sky;
> I pass through the pores of the ocean and shores;
>   I change, but I cannot die.
> For after the rain when with never a stain
>   The pavilion of Heaven is bare,
> And the winds and sunbeams with their convex gleams
>   Build up the blue dome of air,
> I silently laugh at my own cenotaph,
>   And out of the caverns of rain,
> Like a child from the womb, like a ghost from the tomb,
>   I arise and unbuild it again.

The circularity of the water cycle, as invoked by the last stanza, had long been considered one of nature's most harmonious systems. In fact, one of Shelley's sources for 'The Cloud' was a recent translation of a celebrated Sanskrit water poem, the *Meghadūta*, or 'Cloud Messenger', by the fifth-century Hindu poet Kālidāsa, known among the Romantic poets as 'the Shakespeare of India'. The ancient poem narrates the misfortunes of a servant to the Hindu god of wealth, exiled on a mountainside for displeasing his master. Watching the clouds gathering to the south, the servant is moved to ask one of them to waft his sorrows down to the wife he left behind. He then imagines the journey made by the cloud as it moves northwards on its errand of mercy:

> On *Naga Nadi*'s banks thy waters shed,
> And raise the feeble jasmine's languid head:
> Grant for a while thy interposing shroud,
> To where those damsels woo the friendly cloud;
> As while the garland's flowery stores they seek,

The scorching sun-beams singe the tender cheek,
The ear-hung lotus fades, and vain they chase,
Fatigued and faint, the drops that dew the face.[26]

The merciful cloud, having rained down its message of love, then makes its return to the mountains to replenish itself through the agencies of convection and condensation. Shelley's cloud thus mingled recent scientific innovation with ancient poetic ideas, both of which were concerned with the mysteries of atmospheric circulation.

Caspar David Friedrich, *The Wanderer above the Sea of Fog*, 1818, oil on canvas. Friedrich's dramatic painting was more theological than meteorological, taking its cue from Martin Luther's German translation of the Bible, in which 'a fog went up from the Earth and moistened all land' (Genesis 2:6).

Clouds after Howard

In the decades following its publication, Howard's classification underwent a series of modifications and amendments, including the division of the generic clouds into species, such as *cumulus congestus* and *cumulus fractus*, as well as the introduction of the new mid-level clouds *altostratus* and *altocumulus*. By then, the idea of grouping clouds by altitude rather than by shape had gained credence, and in 1887 two leading meteorologists, Professor Hugo H. Hildebrandsson of the University of Uppsala and the Hon. Ralph Abercromby of the Royal Meteorological Society, proposed a new international cloud classification, divided into three altitude bands: low, medium and high clouds. 'The primary idea', they wrote in response to widespread complaints that Howard's classification was being inconsistently used, 'was that the name of a cloud is of far less importance than that the same name should be applied to the same cloud by all observers.'[27] Their idea of a globally agreed classification met with wide support, and in 1890 the first multilingual illustrated cloud atlas was published in Hamburg. In it, Howard's original seven cloud types were expanded to ten: *cirrus, cirrostratus, cirrocumulus, alto-cumulus, altostratus, stratocumulus, nimbus, cumulus, cumulonimbus* and *stratus*.

The international scope of the proposed classification had, in the meantime, raised an obvious question: are clouds the same the world over? The aristocratic Abercromby spent two years

and much of his inherited wealth travelling the world in search of an answer, and on his return published a remarkable memoir entitled *Seas and Skies in Many Latitudes; or, Wanderings in Search of Weather* (1888). There was more than a touch of Phileas Fogg about the intrepid Abercromby, not least his impressively waxed moustache, and some passages of his book were indeed reminiscent of Jules Verne's novels, especially *Five Weeks in a Balloon* (1863), in which the narrator notices 'masses of cloud now piled up under the car. They rolled over one another and formed a confused mass of light as they reflected the rays of the sun.'[28] Yet *Seas and Skies* was a masterclass in sustained observation, at the end of which Abercromby concluded that clouds can be broadly divided into two principle types – heaps and layers – with the former more prevalent in warmer climates, and the latter more common in colder: 'The sky was always more or less clouded near the Equator, mostly with small trade cumulus, leaning forwards from the north-east; replaced farther south by thin flaky stratus, which looked like a venetian blind on the horizon.'[29]

Although Abercromby's method of ship-board observations have been displaced by satellite technology, a recent cloud map of the world released by NASA supports his picture of the relative cloudiness of the equatorial regions. This band of cloudiness is caused by large-scale circulation patterns (known as Hadley cells), in which cool air sinks near the 30-degree latitude line north and south of the equator, while warm air rises near the equator. These restless convergences cause water vapour to condense into cloud, producing a dependable band of thunderclouds in an area known as the Intertropical Convergence Zone.

At any one time, nearly 70 per cent of earth's surface will be covered by clouds, particularly over the oceans, where the phenomenon of upwelling (cooler water from deep in the ocean rising to replace the surface water) creates a layer of cool water at the surface, chilling the air immediately above, forming low stratocumulus clouds. Stratocumulus is the earth's most abundant cloud type, and typically covers around one-fifth of the globe at any one time.

Following the publication of the Hamburg atlas, the International Meteorological Conference convened a 'cloud committee' to oversee the creation of an official global cloud atlas that was published in 1896, which they named the 'International Year of Clouds'. Its appearance marked the conclusive worldwide adoption of Howard's cloud nomenclature, although to compare the original classification of 1802 with the version published in 1896 is to realize how far meteorology had advanced over the course of the nineteenth century. As William Clement Ley pointed out:

> Luke Howard was a very minute and accurate observer; but in his day the laws which regulate the movements of the atmosphere were not understood. The distinctions of cyclone and anticyclone, and the relation of wind and weather to the distribution of barometric pressure were totally unknown.[30]

Clouds were now to be classified firstly by height, in a five-tier arrangement running from 'Upper Clouds' to 'High Fogs', and then within each height category there were two further divisions: '*a*. Separate or globular masses (most frequently seen in dry weather), and *b*. Forms which are widely extended, or completely cover the sky (in wet weather).'[31] These were the 'heaps' and 'layers' into which Abercromby had grouped all clouds during his long meteorological travels. The 1896 cloud ranking ran as follows:

A. Upper Clouds, average altitude 9,000 m
a. 1. *Cirrus*
b. 2. *Cirro-stratus*

B. Intermediate Clouds, between 3,000 m and 7,000 m
a. 3. *Cirro-cumulus*
a. 4. *Alto-cumulus*
b. 5. *Alto-stratus*

C. Lower Clouds, 2,000 m
a. 6. Strato-cumulus
b. 7. Nimbus

D. Clouds of Diurnal Ascending Currents
8. *Cumulus*; apex, 1,800 m; base 1,400 m
9. *Cumulo-nimbus*; apex, 3,000 m to 8,000 m; base, 1,400 m

E. High Fogs, under 1,000 m
10. *Stratus*

The highest-rising cloud of all, cumulonimbus, was numbered nine on the list (as it had been in 1890), and it was not long before the expression 'to be on cloud nine' was coined, meaning 'to be on top of the world' – an apt analogy, given that tropical cumulonimbus clouds can rise to the edge of the stratosphere, some 18 km (30,000 ft) above the earth. Numerical variations of the happy saying have come and gone over the years, including to be on 'cloud seven' (possibly from a conflation with 'seventh heaven'), while Albin Pollock's dictionary of American slang, *The Underworld Speaks* (1935), has an entry for 'cloud eight: befuddled on account of drinking too much liquor'.[32]

Later editions of the *International Cloud Atlas* saw the clouds reordered, with cumulonimbus placed at the end of the list as cloud ten. Recently, however, the World Meteorological Organization, who 'realized they were being spoilsports', renumbered the ten clouds from zero to nine, thereby returning cumulonimbus to its earlier, euphoric status as the true cloud nine: a rare example of science surrendering to symbolism.[33]

During the Second World War, a secret application of the cloud classification was developed for the use of air force pilots. It can be found in the British Air Ministry's *Cloud Atlas for Aviators* (1943), which until quite recently was still a restricted document. The atlas ranked clouds according to their operational value in aerial combat, ranging from 'none' (cirrus) to 'best of all' (altostratus), via 'great' (stratocumulus), 'little value' (altocumulus) and 'dangerous' (cumulonimbus). The ranking was based

Bronze relief by Philip Lindsey Clark from a First World War memorial, Borough High Street, London, *c.* 1922, depicting a dogfight taking place against a background of finely sculpted clouds.

on each cloud's value as a visual obstacle, whether for defence – 'Aircraft may evade a pursuing aeroplane by entering cloud and changing course or changing height when screened by the cloud' – or attack: 'Aircraft may approach a target under cover of the cloud (blind flying) and endeavour to come out of the cloud only when within striking distance.'[34] The publication considered every cloud type in turn: cumulus, for example, 'may be used for evasion. Cover is not continuous, but pilots will usually be able to effect an alteration of course or altitude which will be unobserved by a hostile aircraft'; while altostratus 'is of great importance. It affords continuous cover and pilots will be free to make unobserved alterations of course.'[35] Although the language of meteorology has long been enriched by military allusions – weather fronts were named after the battle zones of the First World War, while drastic drops in atmospheric pressure are known as 'weather bombs' – it is still disquieting to think of clouds being classified (in both senses of the word) as tactical hardware for use against an incoming enemy.

An ascent through the ten cloud types

'A *Cloud* is nothing else, but a great heap of *Snow* close clinging together', as René Descartes observed in 1637, and in a strictly material sense he was right: most clouds consist of ice crystals and supercooled water droplets that have condensed around billions of tiny nuclei present in the atmosphere between 1 and 10 kilometres above the earth's surface.[36] They are shaped by atmospheric factors such as altitude, temperature, humidity, wind shear and the cleanliness of the air (the cleaner the air, the fewer clouds will form), so learning to read their forms over time has become a means to apprehend the otherwise invisible dynamic processes that determine our wider planetary experience of weather and climate.

Cloud formation is largely dependent upon temperature and humidity: the warmer the air, the more water vapour it can hold. When warm air rises through convection or some other form of lifting, it cools until it meets the dew point, the temperature at which water vapour condenses into visible liquid droplets. But condensation – the transition from vapour to liquid – can only happen in the presence of solid or liquid particles known as condensation nuclei: in the case of clouds, these are typically microscopic airborne particles of dust, sea salt or pollen grains, all of which are naturally and abundantly present in the atmosphere. Absolutely clean air has great difficulty condensing its water vapour, and will often supersaturate instead, holding more vapour than is theoretically possible until suitable nuclei turn up.

Once condensation has occurred, the tiny individual water droplets, no more than a millionth of a millimetre across, have enough air resistance to remain suspended in the atmosphere, where, bundled together in their billions, they make up a visible cloud: water vapour itself is not visible, but once condensed, the billions of droplets usually appear white from reflected sunlight. When clouds are particularly thick, the droplets scatter or absorb light instead, allowing less solar radiation to travel through them, which is why storm clouds appear dark grey or even black.

Depending on air temperature and altitude, a cloud may also be formed of billions of microscopic ice crystals alongside the liquid droplets; very cold upper air will freeze all the water molecules into ice, forming the distinctively white, fibrous clouds known as cirrus – the last clouds that James Glaisher would have seen before he passed out in the basket of his balloon.

Let us imagine that we are in one of Glaisher's air balloons, about to ascend (safely) with him to the top of the troposphere, and that we will be lucky enough to pass through all ten cloud types on the way. What would such a journey be like?

Just above the ground, perhaps, will be a misty layer of stratus cloud, which forms in cool, stable conditions, often overnight. Low-level condensation is required to generate this feature-less grey cloud, the base of which will usually be below 500 metres and sometimes low enough to obscure the tops of high buildings. Layers of stratus tend to be broken up by the warmth of a rising sun, or by the arrival of a layer of warm, turbulent air. Goethe, writing in his travel journal in September 1786, described the evaporation of a mountain stratus with close poetic attentiveness:

> I saw quite distinctly the absorption of one such cloud. It clung to the steepest summit, tinted by the afterglow of the setting sun. Slowly, slowly, its edges detached themselves, some fleecy bits were drawn off, lifted high up, and then vanished. Little by little the whole mass disappeared before my eyes, as if it were being spun off from a distaff by an invisible hand.[37]

Watching the behaviour of stratus clouds over time can offer useful indications of short-term future weather: if a layer forms in air rising over hill slopes, it is usually followed by rain, but if low stratus forms during summer nights, the following morning can start off gloomy, but the rising sun will soon evaporate it, leaving the rest of the day fine and clear.

So, as our morning stratus begins to evaporate, we begin our balloon ascent through the lower atmosphere, where we soon encounter a parcel of cumulus clouds, the classic white clouds of a summer sky, of the kind that a child might draw. Cumulus form above thermals, columns of sun-warmed air, whose payloads of water vapour begin to condense at heights of around 600 metres, growing upwards and outwards into the familiar, flat-based 'fair-weather' clouds known as *cumulus humilis*. If these smaller clouds continue to grow, perhaps after rain on a warm afternoon, they begin to be classified as *cumulus mediocris*, which are as tall as they are wide, or even *cumulus congestus*, which are beginning to tower upwards, powered by the release of latent energy during the condensation process. This is Philip Larkin's 'high-builded cloud/ Moving at summer's pace'.[38] The more this cycle continues, the taller the cloud will grow, especially if the process begins early in the morning, with a full day of sunshine ahead of it. The cumuliform cycle depends in large part on the stability and temperature of the surrounding air: if the temperature of the rising moisture matches that of the surrounding air it will tend to spread out, forming stratiform clouds, but if the rising moisture is surrounded by cooler air, it will form taller cumuliform clouds, a sure sign of atmospheric instability. But not all cumulus clouds are destined to grow into larger formations and, as the air begins to cool at the end of a calm summer day, they might start to sink and dissipate, breaking down into fragments of well-named *cumulus fractus*.

Meanwhile, the thermals on which the cumulus clouds formed are helping our balloon to rise through the last of the low clouds, a widespread blanket of stratocumulus. These are the commonest clouds on earth, with a number of different appearances, such as the flat, extensive variety known as *stratocumulus stratiformis*, formed (as its name implies) from the lifting or breaking up of low stratus clouds by means of upward convection. These rising sheets thicken and merge to form a continuous layer of dense grey cloud (a thicker cloud, with more droplets, will absorb light rather than reflect it, though they can also form a pattern of cloudlets with blue sky visible between the clumps). Stratocumulus clouds need to grow thick and dark before they can produce anything but the lightest scattering of drizzle,

Mid-19th-century engraving showing cloud types by height.

although the convective turrets of *stratocumulus castellanus* can develop over a sunny, convective day into cumulus congestus or even rainy cumulonimbus clouds.

A rapidly evaporating cumulus fractus cloud.

We have now ascended above 1,000 metres and into the realm of the mid-level clouds, where we encounter altostratus, a generally featureless layer of thin, grey-blue cloud that can spread to cover most of the sky. Altostratus is usually formed by the lifting of a large mass of air ahead of an incoming warm front; if the warm front continues to advance, pushing further moist air upwards, altostratus can thicken into nimbostratus, a sure sign of imminent rain (its name means 'layer of rain cloud'). Nimbostratus is probably the world's least favourite cloud, being a long-lasting rain-sodden grey blanket that blots out the sun and depresses the spirits. Although it is classified as a medium-level cloud, nimbostratus can form quite high in the sky, but thankfully our balloon pulls us through the rain layer, and into a cold layer of altocumulus, an upper mid-level cloud type that forms either from the breaking up of a sheet of altostratus or from pockets of moist air that are lifted and cooled by gentle turbulence. There are a number of varieties of altocumulus cloud, including the near-stationary wave-cloud *altocumulus lenticularis*, which is commonly mistaken for a UFO, and *altocumulus stratiformis*, the blanket of cloudlets that (like the higher cirrocumulus cloudlets) are more familiarly known as a 'mackerel sky'. *Altocumulus castellanus* cloudlets resemble small turrets or battlements in the sky, and are a sure sign of upper-air instability.

As we ascend further into the cold air, 5 or 6 km higher, we encounter a layer of cirrostratus, high ice-crystal clouds that often form ahead of advancing weather fronts. Their movements are worth keeping an eye on, as they can give a good indication of

Winter sky: snow-laden stratocumulus clouds over southwest Iceland, January 2015.

weather to come: if they thicken and spread across the sky, for instance, that is a sure sign of imminent rain. Just above them are some wispy cirrus clouds, whose diffuse appearance is due to the lower concentrations of ice crystals compared to the higher droplet concentrations found in liquid water clouds. They come in a variety of forms, from *cirrus uncinus*, the easily recognized 'hook' clouds that Luke Howard described as 'pencilled, as it were, across the sky', to great sky-filling displays of *cirrus fibratus* arranged into parallel bands as far as the eye can see. These, too, are worth keeping an eye on, for if they thicken and spread, covering the sky, bad weather is sure to follow.

Our penultimate cloud is cirrocumulus, among the rarest of all cloud types, which forms high in the atmosphere – sometimes as high as 14 km (8.5 miles) – from a mixture of ice crystals and supercooled water droplets. Cirrocumulus often form when convective currents encounter high cirrus or cirrostratus clouds, converting some of their ice crystals into supercooled droplets and breaking them down into grainy ripples of cloud. Due to the unstable conditions in which they form, cirrocumulus tend

Layer of altocumulus cloud lit by the setting sun.

to be short-lived, either thinning out into veils of cirrostratus, or joining up with neighbouring clouds to form a shallow, continuous formation across the sky, the rippled variety of which is known as a 'mackerel sky', a harbinger of stormy weather to come. The old seafarers' saying, 'mare's tails and mackerel scales make tall ships carry low sails', is a testament to the fact that a large amount of moisture borne high in a cold sky is a visible indication of an advancing depression.

We have come to our final tropospheric cloud. As was noted earlier, cumulus clouds can grow to enormous heights, and while our balloon has been rising, we have been keeping an eye on those earlier cumulus congestus clouds, whose rapid, bubbling

'Mackerel sky' of alto-cumulus cloudlets over Edmonton, Alberta.

ascent has mirrored our own; they have now grown massive and unstable through powerful upward convection, and as their summits begin to lose their characteristic cauliflower appearance, they take on the striated look of the full-blown *cumulonimbus capillatus* cloud, the towering thunderhead whose crown spreads out against the tropopause (the boundary between the troposphere and the stratosphere, nearly 20 km (12 miles) above the earth), creating the characteristic anvil formation. These vast roiling structures can weigh 1,000,000 tonnes and contain as much thermal energy as ten atomic bombs: they are the clouds responsible for the noisy drama of an electrical storm, the engines of thunder and lightning. Lightning is formed when clusters of positive and negative charges become separated by powerful updraughts of air within the cloud, congregating in different regions: strong positive charges often build up at the top, while negative charges cluster at the base. When the

This image, taken from the ill-fated *Challenger* space shuttle in 1984, shows the spreading anvils of a group of cumulonimbus storm clouds over Brazil. The spreading occurs as the tops of the clouds meet the tropopause (the boundary between the troposphere and the stratosphere).

Icy cirrus clouds sculpted by the freezing winds, some 6 km (20,000 ft) up.

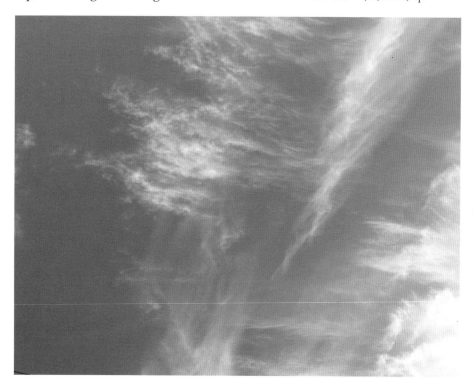

potential difference between these charged regions becomes too great to bear, electrical energy is discharged in the visible (and auditory) form of thunder and lightning.

There are three main varieties of lightning. Cloud-to-ground lightning (often described as forked lightning) occurs when the negative charges in the base of the cloud induce opposite charges in the ground below, sending a powerful spark flying between the differently charged regions. In-cloud (or sheet) lightning happens when sparks fly between differently charged regions within an individual cloud, and thus do not reach the ground. Sparks that fly across the air from one charged cloud to another, by contrast, are known as cloud-to-cloud lightning. Thunder, meanwhile, is the sound created by the sudden expansion of the air column as the lightning surges through it at temperatures of up to 28,000°c. Whether one hears a sudden crack or a low rumbling sound depends upon how short the stroke was, as well as how far away one is from the scene of the strike. Thunder can sometimes be heard nearly 30 km (18 miles) away, but over such distances the sound waves will usually have broken up into vague, indistinct rumbles.

So, given its strong electrical potential, we will keep a wide berth as we circle round the cumulonimbus cloud, before slowly beginning our 20-km (12-mile) descent, having ticked off all ten clouds in the official classification.

Accessory clouds and supplementary features

In addition to the ten main clouds that have just been described, there are three accessory clouds, which occur in conjunction with the principal types, as well as six supplementary features, most of which play supporting roles in the epic dramas staged by large cumulonimbus clouds.

The first accessory cloud is *pannus* (from the Latin for 'cloth' or 'rag'), dark, ragged shreds of either stratus fractus or cumulus fractus that appear below other clouds, usually altostratus, nimbostratus, cumulus or cumulonimbus. Pannus clouds have historically been referred to as 'scud' or 'messenger' clouds,

Rainbow and complex
rainclouds fill the
sky above Stockton,
California.

particularly by sailors and farmers, for whom the message they brought was rain.

The second accessory cloud is *pileus* (from the Latin for 'cap'), a layer of flattened cloud that sometimes appears above a cumulus congestus or cumulonimbus cloud. It is formed by the rapid condensation of a layer of moist air that is pushed up over the cloud's main summit, where it freezes into a layer of icy fog. Pileus clouds are often short lived, with the main cloud beneath them rising through convection to absorb them. Cap clouds can also appear over mountains, and have generated numerous local weather sayings, such as this from Worcestershire: 'When Bredon Hill puts on his hat/ Ye men of the vale, beware of that.'[39] Pileus have even been observed to form above volcanic ash clouds, as a parcel of atmospheric water vapour is suddenly lifted by the volume of erupting material, condensing as it meets a layer of colder air.

The third accessory cloud is *velum* ('sail' or 'awning'), a thin, low, long-lasting layer through which the summits of cumulus congestus or cumulonimbus clouds will often pierce. In contrast

to icy, short-lived pileus clouds, velum clouds are formed in layers of stable, humid air that are lifted by the convective currents within the main cumuliform clouds, and they can persist even after their host clouds have dispersed or decayed.

    One of the six supplementary features is *Arcus* ('arch'), a distinct shelf or roll of low cloud that can appear below a powerful cumulonimbus cloud. It is formed by strong downdraughts of cold air that spread out ahead of the oncoming storm cloud, pushing up layers of warm air nearer the ground to form dense, horizontal rolls of cloud.

    Another, *Incus* ('anvil'), refers to the anvil-shaped summit of a large cumulonimbus cloud, an icy canopy that can grow to enormous heights above the main body of the cloud. It spreads out laterally when it meets the tropopause to create the characteristic flattened thunderhead. It can be smooth in appearance,

Cap cloud (pileus) forming above an eruption of Sarychev Peak, a volcano in the Kuril Islands, north Pacific.

especially at a distance, but it is usually highly fibrous and striated, being composed of billions of ice crystals borne aloft by vigorous upward convection.

Mammatus ('udders') are pouch-like protuberances that form beneath the anvils of large cumulonimbus clouds. They are caused by the sudden sinking of pockets of cold air from the upper to the lower parts of the cloud, reversing the usual pattern of summer cloud formation, when warm, moisture-laden air convects upwards. They are usually associated with stormy, unstable conditions, although they can also appear some time after a storm has passed. Mammatus were known in Scotland as a 'pocky cloud', from the dialect word *pock*, meaning 'bag'.

Praecipitatio ('rain'), from the Latin for 'fall', is applied by meteorologists to a cloud from which any kind of rain, snow, sleet or hail manages to reach the ground, as distinct from *virga*,

Veil cloud (velum) pierced by a rising cumulus congestus cloud over the Preseli Hills, Wales.

which refers to rain that evaporates before it hits the ground. Fog and mist, by contrast, are classified as 'suspensions', since the water vapour they contain does not condense sufficiently to precipitate. Precipitation is a major component of the water cycle, and is responsible for depositing most of the fresh water on the planet. Approximately 505,000 cubic km of water fall as precipitation each year, two-thirds of which fall directly into the oceans. Precipitation occurs when small droplets within the cloud coalesce via collision with other raindrops or ice crystals; the process is often triggered by the upward movement of large parcels of moisture arriving with advancing weather fronts. Precipitation is divided into three broad categories, based on whether it falls as liquid water (rain and drizzle), water that freezes on contact with the lower atmosphere (freezing rain or

Mammatus (from the Latin for 'udders') form beneath the anvils of large cumulonimbus clouds, and are caused by the sudden sinking of pockets of cold air from the upper to the lower parts of the cloud.

Heavy shower falling from a rain-darkened cumulonimbus cloud over the Western Isles, Scotland.

freezing drizzle), or ice (snow, ice pellets, hail and the snow-pellets known as graupel).

Rain itself can vary enormously, depending on the kind of cloud from which it falls. Widespread, persistent rain is most likely to fall from a nimbostratus cloud, while sudden, heavy bursts or sharp showers are more likely to come from large cumulus or cumulonimbus clouds. Most rain begins life as snow in the frozen body of a cloud, melting in the warmer air as it descends. In colder weather, it will fall as unmelted snow, or as sleet, depending on whether it has melted and refrozen as it passes through mixed layers of warm and cold air. If melted snowflakes fall through a sub-freezing layer closer to the surface, they can re-freeze as they descend, forming sleety ice pellets. However, if the sub-freezing layer beneath the warm layer is too narrow, the precipitation will not have time to re-freeze, and gelid, freezing rain will fall instead.

Hail, by contrast, is formed by warm updraughts of air hurling falling ice pellets back into the frozen centre of the cloud. The pellets grow through collision and freezing, rising and falling several times, adding new layers of ice, until they are too heavy to stay aloft, at which point they fall to the ground as hail. The heaviest hailstone ever recorded fell in Bangladesh in April 1986, weighing in at an impressive 1.02 kg (2.2 lb).[40]

Tuba ('funnel cloud') sometimes form at the base of a cumulonimbus cloud when a column of swirling air begins to rotate, condensing ambient moisture into water droplets as it does so. This vortex then begins to move downwards, creating a tapered cone or funnel shape that protrudes some distance below the cloud, although it is rarely strong enough to make contact with the ground. When it does so, it usually takes the form of a weak landspout or waterspout rather than a fully fledged tornado, which tends to develop from the large-scale rotation of a tropical

A bank of high altocumulus cloud passes some 5 km (16,500 ft) above the summits of low-lying cumulus clouds, illustrating the layered nature of the troposphere. This photograph was taken on the same day in the same place as the previous image, showing the mercurial nature of the fringes of weather systems.

supercell thunderstorm, in contrast to the weak vorticity of the cold air funnel cloud.

Luke Howard described a funnel cloud that he spotted near the Yorkshire coast, in a letter to his grandson, dated '17th July 1851, noon':

> We have been entertained within this hour past by the appearance of that very rare phenomenon the Water-spout. The clouds were heavy to the West and N. West with Rain behind them; when one pretty much in advance of the shower let down the usual <u>Jelly-bag</u> – as the women very aptly described it – but only about half-way to the earth, and not being over any water, as I conclude, there was no reaction that we could perceive from beneath.[41]

Describing it as a waterspout (''tis Neptune shaking hands with Jove', as he excitedly concluded) seems to have been an extrapolation on his part, as it is clear from his detailed description that what he actually witnessed was a far more modest funnel cloud.

Virga ('fallstreaks'), from the Latin for 'rod', refers to any form of precipitation, whether rain, snow or ice, that evaporates before it reaches the ground. Its failure to carry on all the way down is usually due to its passing through a layer of warmer or drier air, although sometimes atmospheric conditions will change, and the virga will be replaced by full precipitation from the same cloud.

All our weather occurs inside the troposphere, where the complex interactions of clouds and wind, air pressure, humidity and temperature determine the day-to-day local conditions that we experience on the ground as weather. Clouds affect weather in a number of specific ways, most notably in their ability to mediate surface temperature: high, thin clouds tend to absorb heat from the sun's rays and radiate it down towards earth, while low clouds tend to cool the earth by blocking incoming sunlight. The fewer clouds in the sky, the less insulation there is, which is why cloudless days can be hotter than cloudy days, but cloudless nights will be colder than cloudy nights. For this reason, cloudy days tend to have a narrower temperature range than clear days. Rain and snowfall can also affect surface temperatures, while also having significant environmental impacts such as floods and thaws.

Although the summits of cumulonimbus thunderclouds often reach the top of the troposphere, they are prevented from pushing their way through it by a temperature shift at the tropopause, so they spread out beneath it instead. But there are two varieties of non-weather cloud that can form beyond the troposphere. The first of these are *nacreous* ('mother of pearl') clouds, also known as polar stratospheric clouds, which appear high in the atmosphere, some 15–30 km (9–18 miles) above the earth, in latitudes above 50 degrees. They form in temperatures below -80°c, from a mixture of nitric acid and ice crystals sourced from parcels of moist air that are forced up through the tropopause by unusually strong atmospheric oscillations.

Virga (rain that fails to reach the ground) falling from an altocumulus cloud before evaporating in mid-air.

Polar stratospheric clouds (also known as 'nacreous clouds') form high in the earth's upper atmosphere. Although beautiful, these clouds appear to enhance atmospheric ozone depletion.

The likeliest time to see them is during a winter sunrise or sunset, when most of the sky is dark, leaving them lit by the sun from beneath the horizon. Their iridescent pastel colours can be magically beautiful, an effect heightened by their enormous distance from the viewer, though they are nothing like as distant as the other kind of non-weather cloud, *noctilucent* ('night-shining') clouds, which are also known as polar mesospheric clouds. These are the highest clouds in the earth's atmosphere, forming beyond the stratosphere, on the fringes of space, some 80 km (50 miles, or 26,000 ft) or more above the earth. They consist of ice particles seeded on microscopic nuclei, although the precise mechanism of their formation in the cold, dry conditions of the mesosphere is still a matter of debate.

Although they remain rare, noctilucent clouds have been sighted more frequently in recent decades, shining ever more brightly and at ever lower latitudes. Why this should be is not yet known, but according to one hypothesis, noctilucent clouds

are formed from plumes of space shuttle exhaust emitted high in the upper atmosphere, and their increased appearance reflects a proportionate increase in space traffic. Given that noctilucent clouds have only been in evidence since the 1880s, the heyday of the Industrial Revolution, it is possible that they will prove to be largely anthropogenic phenomena. If so, the tangible impact of human activity extends far further into the earth's upper atmosphere than had ever been previously imagined.

Polar mesospheric clouds (also known as 'noctilucent clouds') are the highest clouds in our atmosphere, forming on the fringes of space, some 80 km (50 miles) above the earth. This occurrence appeared above Stockholm, Sweden, in July 2014.

3 The Language and Literature of Clouds

My long poem was called 'Clouding Up' and it went on for two
and a half pages. It was mostly about clouds. There was something
in it about the Cloudboys and the Nimbians, I wince to remember.
It was a poem I'd written in college. My creative writing teacher,
a taciturn but fair-minded man, wrote, 'I'm a bit baffled by this.
To be frank, it's boring.'
Nicholson Baker, *Travelling Sprinkler* (2013)[1]

In 1703, the year of the great November storm, an anonymous
English weather diarist began work on a private cloud classification. The diarist was, by his own estimation, a cloud obsessive,
blessed with a 'cloud-born or Nubigenous Genius; like Ixion
engendered of a cloud, I am ever gazing & as it were Returning
to my womb.'[2] According to the historian Jan Golinski, who
has published the only detailed account of this long-neglected
diary, the extraordinary variability of clouds was taken up as a
linguistic challenge by the unknown author, who was induced
to devise what he called a 'speciall language' in order to describe
them. He attempted to classify them into eight categories, each
based on a visual comparison to some commonplace phenomenon, such as the mixture of oils or fats with water, or the
swirling patterns found in marble. Elsewhere he suggested
metaphorical names for different cloud types, 'combs', 'palm
branches', 'foxes' tails' and the like, or employed visual similes
such as fleece, cobwebs, crêpe, spun wool, or raw or finished silk.
Some of the metaphors, as Golinski notes, piled up like clouds
themselves:

> Atmosphere loaded & varnished with Bulging, dull swelling
> Bas-Relieve clouds bloated & pendulous. I style them ubera
> caeli fecunda: sky-cubbies or udders cloudy; they enclosed
> & stufft ye whole visible Hemisphere in colour like Lead =
> vapours or a tall Fresco ceiling, or marble veined grotto.[3]

This bravura passage appears to be describing mammatus cloud, a term from today's official meteorological lexicon that employs the same bodily metaphor ('udders') as did the anonymous diarist, who may in turn have drawn from a passage in the *Rig Veda* that describes the udders of the cows of Indra dropping their nourishing richness on the earth.

'Remarkable appearance of Clouds on the 13th of April 1801', from Hayman Rooke's *Annual Meteorological Register* (1802).

There have been many such amateur cloud watchers over the centuries, only a few of whom ever published their findings. One who did was Hayman Rooke, a retired infantry captain and keen weather diarist, who kept journals over a twenty-year period, from 1785 to 1805, in which he recorded the daily weather over his Nottinghamshire village. On one afternoon in April 1801, as he waited for his mail to arrive, he noticed a striking display of cloud 'streets' – 'small white clouds in radiated columns' – massing over his garden for a quarter of an hour, which was just long enough for him to sketch them in his journal.[4] His account of the unusual alignment was one of the earliest positive identifications of the wind conditions necessary to create it.

Like all fugitive effects, clouds require the presence of a validating observer. Yet, since no one can see the same cloud twice, there are many more clouds than there are observers. As the Victorian meteorologist William Clement Ley pointed out in 1879, 'cloud observation is, in a very large measure, an incommunicable art' that was hindered rather than helped by a specialist language that left many non-scientists baffled.[5] But alongside the new terminology were the surviving remnants of the older vernacular languages of the skies, many of which were collected and preserved by scholars such as the meteorologist Richard Inwards, whose popular compendium *Weather Lore* (1869) went through many editions over the years and is still in print today. As Inwards pointed out in his introduction, much of the material is a testament to the observational acuity of our ancestors, 'keen observers of the signs of the sky', who bequeathed us an evocative spoken language of the weather. In Scots Gaelic, for example, the term *rionnach maoim* referred to 'the shadows cast on the moorland by clouds moving across the sky on a bright and windy day', while *roarie-bummlers* were 'fast-moving storm clouds' (literally 'noisy blunderers'). *Skub* was a Shetland word for 'hazy clouds driven by the wind', while *wadder-head*, also from Shetland, referred to 'clouds standing in columns or streaks from the horizon upwards'. *Water-cast* was a Suffolk term for a small cumulus cloud, while *urp* was Kentish for cloud, along with *urpy* – cloudy.[6] Altocumulus cloudlets are still called a 'mackerel sky' by the British, *ciel pommelé* by the French, *cielo empedrado* ('cobbled sky') by the Spanish and *cielo a pecorelle* ('lamb-clouds') by the Italians. The original Maori name for New Zealand, *Aotearoa*, translates as 'the Land of the Long White Cloud', a reference to the orogenic clouds that fill the skies with daily drama, while the name of the rain-drenched Indian state of Meghalaya means 'the abode of clouds' in Sanskrit.

The modern word 'cloud' has its origins in the Old English 'clod', or 'clud' (meaning 'rock' or 'hill'), which was sometimes rendered in plural as 'clowdys'. The Old Norse loan word *sk̃y* meant 'cloud', while, curiously, the Middle English 'welkin' meant 'sky', even though 'welkin' was a cognate of the Dutch and High German *wolke*, meaning 'cloud'. It was not until the thirteenth century that the word 'cloud' took on its primarily meteorological meaning ('a visible mass of condensed watery vapour floating in the air at some considerable height above the general surface of the ground', as the *Oxford English Dictionary* currently defines it), although the word has always been associated with amassing or accumulating, a meaning that still holds true for the most recent use of 'cloud' as the name for a distributed data storage system. Tellingly, the OED gives a source as early as 1705 for the phrase 'a cloud of informations'.[7]

Cloud nomenclature

Although the story of modern cloud naming begins with Luke Howard, whom we met in the previous chapter, there had been a number of earlier attempts to create a standardized meteorological language. In the 1660s Robert Hooke, curator of experiments at the newly founded Royal Society of London, proposed 'A Method for Making a History of the Weather', along with detailed guidelines on the presentation of data. Wind force, temperature, barometric pressure and humidity were to be expressed in numerical form, while 'visible appearances of the Sky' were to be described in words:

The word 'waterspout' appears to date from the 18th century, before when these impressive phenomena were known variously as 'typhoons', 'cyclones' or 'whirlwinds at sea'. This engraving, by Louis Le Breton, from Margollé and Zurcher's *Les Météores* (1869), shows a vessel assailed by multiple waterspouts (*trombes* in French) descending from the base of a storm cloud.

William Berryman,
Rain, Nairn Castle,
c. 1808–16; one of
Berryman's many
plein-air watercolour
views of Jamaica.

Here should be observed, whether the Sky be clear or
clouded; and if clouded, after what manner; whether with
high Exhalations or great white Clouds, or dark thick ones.
Whether those Clouds afford Fogs or Mists, or Sleet, or
Rain, or Snow, &c. Whether the under side of those Clouds
be flat or waved and irregular, as I have often seen before
thunder. Which way they drive, whether all one way, or
some one way, some another; and whether any of these be
the same with the Wind that blows below.[8]

But while the instrumental data were easily quantified, the
parameters of the descriptive language proved problematic from
the start. There were too many different 'faces of the sky: they
are so many, that many of them want proper names', as Hooke
tellingly observed.[9] After recognizing that the empirical challenge was largely linguistic, he set about creating a standard
vocabulary for use among his fellow naturalists:

Let *Cleer* signifie a very cleer Sky without any Clouds or
Exhalations: *Checker'd* a cleer Sky, with many great white
round Clouds, such as are very usual in Summer. *Hazy,*

a Sky that looks whitish, by reason of the thickness of the higher parts of the Air, by some Exhalation not formed into Clouds. *Thick*, a Sky more whitened by a greater company of Vapours . . . Let *Hairy* signifie a Sky that hath many small, thin and high Exhalations, which resemble locks of hair, or flakes of Hemp or Flax: whose varieties may be exprest by *straight* or *curv'd*, &c. according to the resemblance they bear. Let *Water'd* signifie a Sky that has many high thin and small Clouds, looking almost like water'd Tabby, called in some places a Mackeril Sky. Let a Sky be called *Waved*, when those Clouds appear much bigger and lower, but much after the same manner. *Cloudy*, when the Sky has many thick dark Clouds. *Lowring*, when the Sky is not very much overcast, but hath also underneath many thick dark Clouds which threaten rain. The signification of *gloomy*, *foggy*, *misty*, *sleeting*, *driving*, *rainy*, *snowy*, reaches or racks *variable*, &c. are well known, they being very commonly used. There may be also several faces of the Sky compounded of two or more of these, which may be intelligibly enough exprest by two or more of these names.[10]

This was the first true attempt to create a calibrated language to track the fleeting appearances of the clouds. But while terms such as 'Checker'd', 'Water'd', 'Cloudy' or 'Lowring' had some general descriptive merit, they lacked the kind of unambiguous precision required to establish an agreed quantitative approach. After a few months the experiment was abandoned, and Hooke's weather questionnaires were never distributed again.

The idea of a calibrated language of clouds would be revived in the following century by the Societas Meteorologica Palatina, a scientific society founded in Mannheim in the early 1780s, specializing in weather research. One of its many projects was a classification of cloud conditions, complete with symbols and Latin abbreviations to aid rapid communication between observers:

Abbreviation or symbol	Specification
A	White clouds
Cin	Grey clouds
N	Dark clouds
l	Orange-yellow clouds
r	Red clouds
t	Thin clouds
sp	Thick clouds
fasc	Streak-like clouds
rup	Rock-like clouds
lact	Clouds of milky appearance
⪉	Layered clouds
〰〰	Gathering clouds

The individual terms and symbols could be used separately or in conjunction: *cin.sp*, for example, referred to dense, grey clouds, while *fasc.l* denoted clouds of a streaky orange.[11] As did Hooke's earlier terminology, these combinations hinted at the central idea of cloud modification. Because clouds merge and separate, rarely staying still for long, any successful classification would need to accommodate the weather's essential changeability. The Mannheim meteorologists came close to establishing just such a schema, even if their homely terminology ('thin'; 'thick'; 'rock-like'; 'milky') remained just as imprecise as Hooke's. Had their society not been disbanded by Napoleon's invading army in 1795, the language of clouds that we speak today might have turned out very different.

In December 1802 Luke Howard proposed his now-universal cloud nomenclature but, unknown to him, another cloud classification had been published earlier that year, by the French natural philosopher Jean-Baptiste Lamarck, author of an annual

A
SCHEME

At one View reprefenting to the Eye the Obfervations of the Weather for a Month.

Dayes of the Month and place of the Sun. Remarkable houfe.	Age and fign of the Moon at Noon.	The Quarters of the Wind and its ftrength.	The Degrees of Heat and Cold.	The Degrees of Dryneſs and Moyſture.	The Degrees of Preſſure.	The Faces or vifible ap-pearances of the Sky.	The Nota-bleſt Effects.	General Deductions to be made after the fide is fitted with Obfervations : As,
4 8 14 II 12.46	27 12 ♉ 9. 46. 4 8 Perigeů. 12	W. 2. 3½ W.SW.1	9 ¾ 12 ½ 16 10 ⅛ 7 ½	2 2 2 2	5 29 1/16 8 9 29 ⅛ 29 ⅜	Clear blew, but yellowiſh in the N. E. Clowded to-ward the S. Checker'd blew.	A great dew. Thunder, far to the South. A very great Tide.	From the laſt quar:of the Moon to the change the weather was ve. ry temperate but cold for the ſea-ſon ; the Wind pretty conſtant
8 15 4 II 6 13. 40 10	28 ♉ 24. 51.	N.W. 3 4 N. 2 1	9 4 8 ½ 7	2 8½ 29 1/16 2 9 2 10 29	A clear Sky all day, but a little chec-ker'd at 4. P.M. at Sun-fet red and hazy.	Not by much fo big a Tide as yeſterday. Thunder in the North.	between N. and W. A little before the laſt great Wind, and till the Wind roſe at its higheſt, the Quickſilver con-
16 II 14. 37	10 N.Moon. S. at 7. 25' A.M. II 10. 8.	1	10	1 10 28 ½	Overcaſt and very lowr-ing.	No dew upon the ground, but very much upon Marble ſtones,	tinued deſcend-ing till it came very low ; after which it began to reaſcend,
	&c.	&c.	&c.	&c. &c.		&c.	&c.	&c,

DI-

weather almanac, the *Annuaires météorologiques*. In the almanac's third issue, for 1802, Lamarck discussed the need for 'the clarification of meteorological phenomena', including a classification of clouds based on his observation that 'clouds have certain general forms which are not at all dependent on chance but on a state of affairs which it would be useful to recognise and determine'.[12] But while Lamarck pre-empted Howard's contention that every cloud could be described through a limited number of terms, his classification ended up presenting them as individual entities rather than connected modifications:

1 *brumeux*	thick fog
2 *en voile*	overcast
3 *en lambeaux*	shreds of clouds
4 *boursouflés*	puffs of clouds
5 *en barres*	barred clouds
6 *en balayures*	thin bars of clouds
7 *pommelés*	small dappled clouds
8 *attroupes*	flocks of clouds
9 *coureurs*	running clouds
10 *groupés*	grouped clouds
11 *de tonnerre*	thunder clouds

Like the earlier cloud terminologies, Lamarck's vocabulary lacked precision, with vernacular terms such as 'puffs' or 'flocks' being too approximate for an agreed scientific usage. Even after he began to introduce secondary terms such as 'isolated' or 'undulatory', Lamarck's descriptions remained too general, a quality compounded by his choice of French rather than universal Latin, which was still the primary language of European science.[13] Given that France's war-ravaged neighbours were unlikely to accept a francophone taxonomy during the height of the Napoleonic campaigns, the choice of a local language for an international classification seems astonishingly short-sighted. But Lamarck's suggestion that clouds should be graded by altitude fared better than his proposed terminology, and by the end of the century the international meteorological

Robert Hooke, 'A Scheme at one View representing to the Eye the Observations of the Weather for a Month', from his *A Method for Making a History of the Weather* (1667).

community had adopted Lamarck's tripartite altitude structure alongside Howard's Latin names.

Clouds and other weather phenomena, from J. G. Heck's *Iconographic Encyclopædia of Science, Literature, and Art* (1851).

Competing cloud names

Among the most diligent early users of Luke Howard's new cloud classification was the whaler and explorer William Scoresby, who voyaged to the Arctic in the summer of 1810. Encouraged by his tutors at Edinburgh University, he made detailed meteorological and oceanographic observations, and was the first seaman to adopt the new classification in his ship's log. Arctic clouds, he wrote, 'generally consist of a dense stratum of obscurity, covering the whole expanse of the heavens'. But sometimes:

the cirrus, cirro-cumulus, and cirro-stratus, of Howard's nomenclature, are occasionally distinct; the nimbus is

Cloud images from J. G. Heck's *Iconographic Encyclopædia of Science, Literature, and Art* (1851).

partly formed, but never complete; and the grandeur of the cumulus or thunder-cloud is never seen, unless it be on the land . . . the most common definable cloud seen at sea, is a particular modification, somewhat resembling the cirro-stratus, consisting of large patches of cloud arranged in horizontal strata, and enlightened by the sun on one edge of each stratum.[14]

Scoresby's account was a striking mix of observation and calibration, although anyone reading it would have needed a secure grasp of the new terminology, a specialist language being put to descriptive work. But while Howard's new vocabulary began to be taken up, it was not universally adopted, and many of the older cloud terminologies remained in use for decades to come. When the agricultural journalist William Cobbett encountered a rising bank of cumulus congestus while riding through rural

Surrey in August 1823, for example, his written account made playful use of a long-established visual metaphor:

> I saw, beginning to poke up over the South Downs (then right before me) several parcels of those *white, curled clouds*, that we call *Judges' Wigs*. And they are just like Judges' wigs. Not the *parson-like* things which the Judges wear when they have to listen to the dull wrangling and duller jests of the lawyers; but those *big* wigs which hang down about their shoulders, when they are about to tell you a little of *their intentions*, and their very looks say, '*Stand clear!*' These clouds (if rising from the South West) hold precisely the same language to the great-coatless traveller. Rain is *sure* to follow them.[15]

Cobbett's vividly drawn rain clouds spoke in exactly the kind of picturesque language that Howard's Latin nomenclature had been designed to dislodge. But while professional meteorologists exchanged Howard's newly minted words between themselves, some continued to use older cloud names when writing for a non-scientific readership. In 1859 Robert FitzRoy, the first director of the Meteorological Office (and a diligent promoter of Howard's classification), produced a short weather primer for use at sea, in which he employed an impressive variety of deliberately non-specialist terms:

> Soft-looking, or delicate clouds foretell fine weather, with moderate or light breezes; hard-edged oily-looking clouds, wind . . . Generally, the *softer* clouds look, the less wind (but perhaps more rain) may be expected; and the harder, more 'greasy' rolled, tufted, or ragged, the stronger the coming wind will prove.[16]

Like Cobbett's 'wigs', FitzRoy's choice of terms reached back into long-established forms of weather language with which mariners were more likely to be familiar than Howard's sonorous Latin. But the longevity of these older terms was neither

Steaming sea ice off the Alaskan coast, December 2012.

surprising nor unwelcome, for there will always be a need for different linguistic registers in which to communicate in different ways. Just as few non-zoologists would refer to a goldfinch as *Carduelis carduelis*, few non-meteorologists would call an approaching storm cloud a *cumulonimbus capillatus*. As Desmond King-Hele pointed out in his reading of Shelley's 'The Cloud', 'If an Englishman greeted his neighbour with a bold "Good morning, fine altocumulus castellanus to-day", the ice which had taken years to break would quickly re-form.'[17] Unwelcome, however, were hostile challenges to the new cloud language, especially those made on unscientific grounds. The first of these came in the course of an otherwise favourable notice of Howard's essay in the *Annual Review* for 1804. The reviewer, a Liverpool doctor named John Bostock, praised 'the method pursued by Mr

Howard', but wondered why he had not used 'plain English names in a science, the further improvement of which will, probably, in a considerable degree, depend upon the observations of the unlearned'.[18] It was a question to which Bostock soon returned, publishing a lengthy article in a leading scientific journal, in which he argued that Howard's terminology was 'too confined to be of any great use', while proposing his own alternative nomenclature instead:

Arc: a body of clouds, stretching in nearly parallel lines over a considerable part of the heavens, and converging in a point in the horizon

Linear arc: long parallel lines or threads

Mottled arc: small rounded clouds, lying side by side or in rows

A wreathed arc: resembling a volume of smoke, as it rises from a chimney top

A feathered arc: resembling feathers, having a linear centre and lateral branches

Shaded clouds: when the clouds are formed into rounded masses of greater or lesser extent, one side of which is very much darker than the other side

Piled clouds and rolling clouds: large rounded clouds, which appear as if they were heaped and rolled one upon another

Tufts: clouds which resemble bunches of hair, the fibres of which are sometimes disposed in a perfectly irregular manner

Flocks: when clouds form larger and compact masses than those which I have called *tufts*[19]

Bostock conceded that his terminology might be considered 'uncouth', but maintained nevertheless that an English-language version was preferable to Howard's elitist Latin. A few weeks later, Howard published a detailed response, claiming that in the interval since its original publication he had seen no reason to amend his own classification, while describing Bostock's rival terms as 'inaccurate and imperfect':

Round clouds, shaded clouds, piled clouds, rolling clouds, white clouds on a gray ground, &c. I do not see the advantage to science of these attempts to substitute description for definition. The *piled* cloud will be the *shaded*, or some other, when it comes toward the zenith; and the *shaded* will be light in the horizon opposed to the sun. As for the *rolling* cloud, I have not yet detected it; and it seems too poetical, if, as I conjecture, it is so named because its parts, if solid, would roll when on an inclined plane.[20]

Howard's distinction – between description and definition – was an insightful one, as was his argument that the Latin terms were clearer than Bostock's because they were easier to organize. To object to the use of Latin or Greek terminology, he pointed out, would mean objecting to thousands of loan words, from *alphabet* to *zenith*, that are long-established features of the English language. After this, nothing more was heard of Bostock and his vernacular clouds, though his would not be the last attempt to replace the Latin nomenclature with English-language terms.

In the meantime, a leading science writer, Thomas Forster, had begun the process of adding species and varieties to Howard's generic clouds, with Latin names 'calculated to express the particular shape, figure, or manner of arrangement, which are circumstances quite different from their modifications'. Forster's suggested terms included:

Cirrus comoides ('from its appearing like a distended lock of hair')
Cirrus linearis ('straight lines')
Cirrus filiformis ('a confused bundle of threads')
Cirrus reticularis ('a beautiful network, consisting of light transverse bars or streaks')
Cirrostratus striatus ('composed of long parallel bars')
Cirrostratus undulatus ('finely undulated')
Cirrostratus myoides ('gives the idea of the fibres of muscle')
Cirrostratus planus ('a large continuous sheet')
Cumulus petroides ('rocklike and mountainous')

Cumulus tuberculatus ('numerous roundish tubercles')
Cumulus floccosus ('divided into loose fleeces')[21]

Engraving from Thomas Forster's *Researches About Atmospheric Phænomena* (1813), illustrating what he terms 'Cumulus floccosus' and 'Cirrus comoides'.

Although these proposed supplementary terms had little impact at the time, they closely resemble some of the varietal terms that are in current meteorological use (see Appendix). But if such a proliferation of scientific Latin might have been welcomed by specialists, it was deeply unpopular among a wider public who still struggled with the unfamiliar language. This was brought home to Forster himself when a number of his friends admitted

that they could never remember the technical terms, which were made up of Latin or Greek words, which they did not understand, and wished that names could be given to Meteorological Phænomena, which are formed out of our own tongue.[22]

Inspired by these comments, he set about translating Howard's nomenclature into a curious form of Old English vernacular:

CURL-CLOUD. The old name in Latin by Mr Howard, is Cirrus, a curl; Cirrulus and curl being the diminutive.

STACKEN-CLOUD, or Cumulus, from the verb to stack, to heap up.

FALL-CLOUD, or Stratus; being the falling, or subsidence of watery particles in the evening.

SONDER-CLOUD, or Cirrocumulus, is a sundered cloud, made up of separated orbs. The characteristic of this cloud being the gathering together into a bed of little clouds, yet so far asunder as not to touch.

WANE-CLOUD, or Cirrostratus; from the waning or subsiding state of this cloud in all its forms.

TWAIN-CLOUD, or Cumulostratus; made often by the twining or uniting of two clouds together.

RAIN CLOUD, or Nimbus, speaks for itself. So we can have *Storm cloud*, *Thunder-cloud*, &c.[23]

Forster was pleased with his translations, and sent them to the editors of the *Encyclopaedia Britannica*, urging them to include them in their next edition. His argument − that every animal, mineral and vegetable has a local as well as a Latin name, so why shouldn't clouds? − was persuasive, and in 1824 his translations duly appeared in the encyclopaedia's sixth edition. By this time Forster was publishing them widely in textbooks and almanacs, writing well-observed descriptions of clouds and skies that featured no Latin names at all: 'The change from Curlcloud to Wanecloud [by the 1820s, Forster writes Curlcloud and Wanecloud without hyphens], and indeed the great prevalence of the latter cloud at any time, must be regarded as an indication of an impending fall'; or, 'We have seen little Stackenclouds form and disappear in the space of a few minutes; while the Curclouds form, change their figures to spots of Sondercloud, and disappear.'[24]

For Howard, this was an unwelcome development, and in his introduction to *The Climate of London* (1818), his pioneering work of urban meteorology, he outlined his objections to Forster's translations:

I mention these in order to have the opportunity of saying that I do not adopt them. The names for the clouds which I deduced from the Latin are but seven in number, and very easy to remember: they were intended as *arbitrary terms* for the *structure* of clouds, and the meaning of each was carefully fixed by a definition: the observer having once made himself master of this, was able to apply the terms with correctness, after a little experience, to the subject under all its varieties of form, colour, or position. The new names ... are superfluous; if intended to convey in themselves an explanation in English, they fail in this, by applying only to some part or circumstance of the definition; the whole of which must be kept in view to study the subject with success. To take for example the first of the modifications – the term *Cirrus* very readily takes an abstract meaning, equally applicable to the rectilinear as to the flexuous forms of the subject. But the name of *Curl-cloud* will not, without some violence to its *obvious* sense, acquire this more extensive one; and will, therefore, be apt to mislead the learner, rather than forward his progress.[25]

Here was the same distinction between definition and description that Howard had made with reference to Bostock's English versions, reinforced by the defence of Latin as 'an universal language, by means of which the intelligent of every country may convey to each other their ideas without the necessity of translation'. This was a powerful line of defence, and by the time Forster's Anglophone cloud names had been dropped from the seventh edition of the *Encyclopaedia Britannica* (1842), they had been all but forgotten in favour of the Latin terminology.

By then, however, Latin itself had begun to decline as the international language of scholarship, with most scientific publications appearing in modern vernaculars. In nineteenth-century Germany, as in many other European countries, Latin had become so unfamiliar to all but a learned minority that even terms such as 'cirrus' and 'stratus' struck an awkwardly classical note. So, while Forster was translating the cloud terms into

English, a physics professor in the university at Halle was doing the same in German. When Ludwig Wilhelm Gilbert published his translation of Howard's essay in 1815, cirrus was rendered as *Die Locken- oder Feder-Wolke* ('hair or feather cloud'), cumulus as *Die Haufen-Wolken* ('heap cloud'), stratus as *Die Nebel-schicht* ('fog layer'), and cirrocumulus as *Die Schafwölkchen* ('sheep cloudlets').[26] But unlike Forster, Gilbert had reservations about the value of such vernacular translations, and conceded that 'cirrostratus' had proved so difficult to translate that he had left it in the original Latin.

The poet Goethe, who had not only praised but versified Howard's cloud classification, viewed Gilbert's efforts as misguided, arguing that 'the names should be accepted in all languages; they should not be translated':

> By this patriotic purism of style nothing is gained; for since it is known, as it is, that clouds only are being discussed, it does not sound well to speak of 'Heap Clouds,' &c. and to keep repeating the general when speaking of the particular. In other scientific descriptions this is expressly forbidden.[27]

But Goethe's distinction between the general and the particular was a familiar one, and did little to stem the enthusiasm for meteorological translations. In the 1860s the Cuban meteorologist Andrés Poey, of the Havana Observatory, called for a complete restructuring of the cloud classification to include an enlarged Latin nomenclature, a system of convenient symbols for use in observatory logbooks, and corresponding vernacular terms in French, English, German, Italian and Spanish in order to ensure that observers could match their local experiences with the agreed scientific categories. The idea was trialled in the first multilingual *Cloud Atlas* (1890), which featured French, German and English translations of the Latin terms, such as 'feather-cloud' (cirrus), 'veil cloud' (cirrostratus), 'woolpack-cloud' (cumulus), 'fleecy cloud' (cirrocumulus) and 'lifted fog' (stratus), although the editors of the first official *International Cloud Atlas*

(1896) dropped these curiously pastoral translations in favour of Latin only.

By then the classification had been modified and amended by several hands – a process that continues to this day. The first, minor, amendment to be made was the reordering of Howard's term 'cumulo-stratus' to 'stratocumulus', as a way of re-categorizing this low-pressure cloud from the convective cumulus family into the more appropriate stratiform clouds. As was seen in the previous chapter, prefixes are an important element of cloud classification, and 'cumulo-stratus' gave a misleading impression of this largely stratiform cloud type that Ralph Abercromby characterized as 'not flat enough to be called pure stratus, but rising into lumps too irregular to be called true cumulus'.[28]

Andrés Poey had also proposed two new cloud species to be added to 'la belle classification', which he termed *pallium* (from the Latin for 'cloth') and *fractus* (from the Latin for 'broken'). These would be used to designate specific appearances of already existing clouds, thus *fracto-cumulus* would describe small, broken fragments of cumulus cloud, while *pallio-cirrus* would denote widespread clouds with what Poey luminously described as *'l'apparence d'un manteaux ou d'un voile d'une dimension considerable, d'une texture très-serrée'* ('like a cloak or cloth').[29] Confusingly, Poey went on to develop a number of objections to the existing classification, arguing that cumulus is 'not a distinct species of cloud' and stratus is 'not a cloud properly so called, but only a mist or hoar frost', but his views failed to attract meteorological support. Just as confusingly, William Clement Ley attempted to introduce his own entirely redrawn classification, based on the formation processes of clouds, subsuming the existing terminology into four newly identified categories that he termed 'clouds of radiation' (fog or low stratus); 'clouds of interfret' (wave-clouds produced at the interface of two or more air-streams); 'clouds of inversion' (rapidly growing cumuliform clouds that rise through the layers of the troposphere); and 'clouds of inclination' (cirriform clouds that occur in the higher, colder regions of the atmosphere). Proposed names for a battery of new cloud species, including *Stratus quietus* ('quiet cloud'),

'The Glorious Picture Gallery of the Sky': a page of early cloud photographs from *The Book of Knowledge* (1926), a popular family encyclopedia.

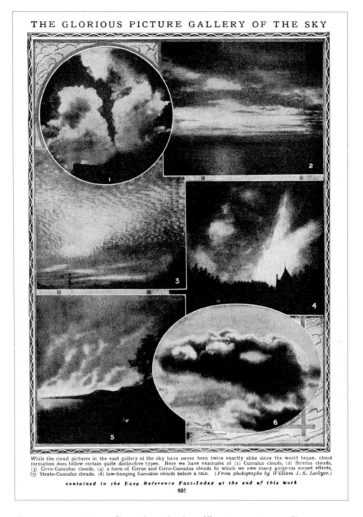

THE GLORIOUS PICTURE GALLERY OF THE SKY

While the cloud pictures in the vast gallery of the sky have never been twice exactly alike since the world began, cloud formation does follow certain quite distinctive types. Here we have examples of (1) Cumulus clouds, (2) Stratus clouds, (3) Cirro-Cumulus clouds, (4) a form of Cirrus and Cirro-Cumulus clouds to which we owe many gorgeous sunset effects, (5) Strato-Cumulus clouds, (6) low-hanging Cumulus clouds before a rain. (*From photographs by William J. S. Lockyer.*)

contained in the Easy Reference Fact-Index at the end of this work

891

Stratus maculosus ('mackerel cloud'), *Cirro-filum* ('gossamer cloud') and *Nimbus grandineus* ('hail shower'), were also explained in great detail in Ley's eccentric volume *Cloudland* (1894), which was distinguished by the use of intricately designed cloud symbols, whose direct descendants can still be found on synoptic weather charts today.[30]

In the end, the original Latin-based nomenclature prevailed, although many people continued to find the terms difficult to remember. In a scene from his last, unfinished, novel, *Bouvard*

et Pécuchet (published posthumously in 1881), Gustave Flaubert pictured his comic antiheroes struggling to fit the names to the passing clouds:

'Typical Varieties of Clouds', from William Clement Ley's 1878 lecture 'Clouds and Weather Signs', reprinted in *Modern Meteorology* (1879).

To get to know something about weather signs they studied the clouds according to Luke Howard's classification. They stared at the ones which stuck out like mares' tails, those that look like islands, those that one might take for snow mountains, trying to distinguish nimbus from cirrus, stratus from cumulus, but the shapes changed before they could find the names.[31]

And the elegant Latin of the clouds continues to present linguistic pitfalls. Keith, the overbearing husband in Mike Leigh's television play *Nuts in May* (1976), muses on the view from Corfe Castle, Dorset, with 'the great nimbocumulus rising above it all like great puffs of cotton wool', while the Conservative politician Alan Clark, in a diary entry written in July 1990, described 'a towering thunderhead of Alto-Cumulus, precursor of change not just in the weather, but in the Climate'.[32] Political metaphors (and random capitalizations) aside, Clark had in fact mistaken small, fleecy altocumulus for large, thundery cumulonimbus, the same cloud word on which the equally pompous Keith had tripped.

Cloud poetry

In Alexander Pope's *Peri Bathous* (1727), an unsparing satirical essay on incompetent writing, he shares a recipe for an epic poem in which the following ingredients should be used:

> For *a tempest*. Take Eurus, Zephyr, Auster and Boreas, and cast them together in one verse. Add to these of rain, lightning and thunder (the loudest you can) *quantum sufficit*. Mix your clouds and billows well together till they foam, and thicken your description here and there with a quicksand. Brew your tempest well in your head, before you set it a blowing.[33]

As Pope's sarcastic advice implies, weather invocation is a particular weakness of poets, for whom the formlessness and evanescence of clouds and skies offer limitless poetic possibilities. For Shelley, whose long-standing interest in meteorology was reflected in a number of major works, including 'Ode to the West Wind' (1819) and 'The Cloud' (1820), clouds were a ready symbol of the inevitability of change and decay, as expressed in the opening lines of his earlier poem 'Mutability' (1814), best known today for being quoted by Victor Frankenstein in the course of Mary Shelley's novel:

> We are as clouds that veil the midnight moon;
> How restlessly they speed, and gleam, and quiver,
> Streaking the darkness radiantly! – yet soon
> Night closes round, and they are lost forever.[34]

Though they are not 'lost' so much as transformed, since for Shelley, as for Goethe, nothing in nature dies or disappears, it only changes ('I change, but I cannot die', as the final stanza of 'The Cloud' declared). Mutability, the disposition to change that characterizes all natural forms, is a particular quality of clouds, whose improvised identities are drawn from processes rather than permanence, from the unceasing transformational energy that makes one cloud morph into another. 'I take great delight in watching the changes of the atmosphere', Shelley declared in a letter sent from Tuscany in July 1818, much of which reads like a projected synopsis of 'The Cloud':

> The atmosphere here, unlike that of the rest of Italy, is diversified with clouds, which grow in the middle of the day, and sometimes bring thunder and lightning, and hail about the size of a pigeon's egg, and decrease towards the evening, leaving only those finely woven webs of vapour which we see in English skies, and flocks of fleecy and slowly moving clouds, which all vanish before sunset.[35]

For Shelley, clouds were recurring symbols of impermanence and loss, while for Elizabeth Barrett Browning, writing a generation later, they promised a refuge for poetic reverie – as invoked in 'The House of Clouds' (1841), a long, nubilous fantasy of aerial retreat that, like Shelley's clouds of mutability, begins with an image of celestial building:

> I.
> I would build a cloudy House
>   For my thoughts to live in,
> When for earth too fancy-loose
>   And too low for Heaven.
> Hush! I talk my dream aloud;
>   I build it bright to see, –
> I build it on the moonlit cloud
>   To which I looked with *thee*.

> X.
> Bring a grey cloud from the east
>   Where the lark is singing,

(Something of the song at least,
　Unlost in the bringing.)
That shall be a morning chair,
　Poet-dream may sit in,
When it leans out on the air,
　Unrhymed and unwritten.

It ends, however, in elemental dissipation:

XII.
Poet's thought, – not poet's sigh.
　'Las, they come together!
Cloudy walls divide and fly,
　As in April weather!
Cupola and column proud,
　Structure bright to see,
Gone! except that moonlit cloud,
　To which I looked with *thee*.[36]

The poem was written in Torquay in the summer of 1841, where Barrett Browning was recuperating following the death of her brother Edward. Her father, she wrote, had liked the poem 'so much that he would wish to live in it if it were not for the damp', though not all who read it responded so favourably to what Browning herself referred to as 'my castle in the air . . . no fit lodging for the critics'. As she wrote to her friend and fellow poet Mary Russell Mitford, who considered 'The Cloud House', as she called it, to be 'one of the most beautiful poems in the language', Browning's tutor, Hugh Stuart Boyd, had been dismayed by the poem, seeing in its misty verses 'the influence of disease on the intellect'.[37]

What is it about clouds that draws the attention of poets? Perhaps, as Alexandra Harris suggests, it is their air of provisionality: 'Like a much redrafted poem there is no single authoritative version of a cloud. The cloud-form is constantly revised and never finished.'[38] This was certainly true for Elizabeth Barrett Browning, who tinkered with 'The House of Clouds' for years, adding new stanzas, removing superfluities, adjusting its tone and tempo as the mood took her, while for the narrator of Nicholson Baker's novel *Travelling Sprinkler* (2013), who winced to remember his early poem 'Clouding Up', the very ungrasp-ability of clouds retained a fatal, bathetic attraction:

> I've written three more poems about clouds since then. I can't get enough of them. I am drawn to describe them even though I know it's futile. They're different every day. Debussy liked clouds. The first movement of his *Nocturnes* is called 'Nuages'.[39]

As noted in the previous chapter, Luke Howard's life as well as his language became a subject for poetry, which, perhaps sur-prisingly, continues to be the case. Lavinia Greenlaw addressed Howard directly in 'What We Can See of the Sky Has Fallen: Luke Howard 1772–1864' (1997), written while she was poet in residence at the Science Museum, London. As had Goethe nearly two centuries before, she invoked Howard's landmark work:

> . . . your essay on clouds: cool distillations
> from your observations' heat. Not giving shape, you found it
> and found yourself ever after skybound . . .[40]

Greenlaw returned to the topic of Howard and his clouds in 2013, composing a new poem for a BBC radio programme, 'The Namer of Clouds', in which she likened the task of poetry to that of meteorology, observing that both are engaged in 'describ-ing imprecision precisely', with Howard's poetic nomenclature providing 'a set of parameters rather than definitions'.[41]

The British Poet Laureate Carol Ann Duffy explored simi-lar territory in her poem 'Luke Howard, Namer of Clouds' (2011), in which the 'smitten' young man loved clouds above all, and set about naming them after everyday objects: 'even a curl of hair – thus, Cirrus./ Cumulus. Stratus. Nimbus.'[42] Even in list form, Howard's scientific Latin forms an evocative narrative, as it does in Billy Collins's 'Student of Clouds' (1991), with its rich pedagogical litany drawn from the language of classification, framing them in the sky – 'long enough to tag them with their Latin names./ Cirrus, nimbus, stratocumulus' – and as it also does in Lewis Grassic Gibbon's novel *Cloud Howe* (1933), which was divided into four parts: 'Cirrus', 'Cumulus', 'Stratus' and 'Nimbus'. The novel – the second instalment of Gibbon's 'A Scots Quair' trilogy – tells the story of Chris Guthrie, a determined heroine torn between her love of the land and a desire to escape the harshness of peasant life. The cloud motifs analogize the movements of the narrative, from the delicate hope of 'Cirrus' to the tragic rain clouds of 'Nimbus', although as Tom Crawford notes in his introduction to a recent reprint of the novel, nimbus carries the secondary meaning of 'a cloud or luminous mist investing a god or goddess', like the halo (or

The low-angled rays of the setting sun illuminate an isolated bank of altocumulus cloud.

'glory') attending a Brocken spectre.[43] Clouds and cloud imagery haunt the increasingly bleak novel, until the final paragraph sees the horizon at last begin to clear:

> She went slow down the brae, only looked back at the frown of the hills, and caught her breath at the sight they held, seeing them bare of their clouds for once, the pillars of mist that aye crowned their heights, all but a faint wisp vanishing south, and the bare, still rocks upturned to the sky.[44]

Asperitas

As was hinted at earlier in the chapter, the process of cloud naming is far from finished. In 2009 Gavin Pretor-Pinney, the founder of the Cloud Appreciation Society, claimed to have identified a new cloud variety, which he named *asperatus*, from the Latin for 'roughened'. Virgil had used the word in the *Aeneid* to describe the surface of a wind-whipped sea, which is what these dramatic wave-like cloud formations most closely resemble. Having had a number of photographs of similarly turbulent clouds submitted to the Cloud Appreciation Society gallery, mostly from the western United States, Pretor-Pinney approached the Royal Meteorological Society (RMS) to enquire about the process of having a new descriptive term added to the official lexicon. After a long period of consultation the RMS approached the World Meteorological Organization (WMO) in Geneva, the intergovernmental organization that oversees the official cloud nomenclature, with a recommendation on behalf of the proposed term 'asperatus'.

After an even longer period of consultation, the WMO announced that it had agreed to recognize the new cloud term, but were categorizing it as a supplementary feature rather than a true cloud variety. They had therefore changed the name to *asperitas* ('roughness'), since the amended classification required a noun rather than an adjective. The official WMO definition of the term is now:

An example of the newly named asperitas cloud variety over Tallinn, Estonia, 2009.

A formation made up of well-defined, wavelike structures in the underside of the cloud, more chaotic and with less horizontal organization than undulatus. It is characterised by localized waves in the cloud base, either smooth or dappled with smaller features, sometimes descending into sharp points, as if viewing a roughened sea surface from below. Varying levels of illumination and thickness of cloud can lead to dramatic visual effects.[45]

Along with two other proposed terms, *volutus* ('rolled') and *homogenitus* ('man-made'), asperitas will be included in the next edition of the *International Cloud Atlas*, to be published in 2017. They will be the first new terms to be added to the official cloud

A sky-filling display of altocumulus stratiformis passing over Nelson's Column, London.

nomenclature since *intortus* ('tangled') was accepted by the WMO more than sixty years ago. And as meteorology continues to evolve and refine, so will its poetic nomenclature, with many as yet unguessable terms likely to be added, over the coming years, to the stained-glass language of the sky.

4 Clouds in Art, Photography and Music

> I'd finally do something I had in mind for years. I'd make a series
> of cloud pictures. I wanted to photograph clouds to find out what
> I had learned in forty years about photography. Through clouds to
> put down my philosophy of life – to show that my photographs
> were not due to subject matter – not to special trees, or faces, or
> interiors, to special privileges – clouds were there for everyone –
> no tax as yet on them – free.
>
> Alfred Stieglitz, 'How I Came to Photograph Clouds' (1923)[1]

With their fleeting lifespans and complex topographies, clouds
have always offered a challenge to representation. When the
eighteenth-century amateur painter (and codifier of the picturesque) William Gilpin encountered a dazzling evening sky while
travelling through Somerset in the 1770s, the prospect naturally

> invited the pencil; but it was a transitory scene. As we stood
> gazing at it, the sun sunk below the cloud, and being
> stripped of all its splendor by the haziness of the atmosphere,
> fell, like a ball of fire, into the horizon.

And before Gilpin could even begin his sketch, he wrote, 'The
whole radiant vision faded away.'[2] Such skies and clouds can be
read as the embodiment of the vanishing moment that flees from
any attempt at scrutiny, clouds being nothing if not elusive, as
Jonathan Swift memorably complained in his pessimistic satire
A Tale of a Tub (1704):

> If I should venture, in a windy day, to affirm to your
> Highness that there is a large cloud near the *horizon*
> in the form of a *bear*, another in the *zenith* with the
> head of an *ass*, a third to the westward with claws like
> a *dragon*; and your Highness should in a few minutes
> think fit to examine the truth, it is certain they would
> be all changed in figure and position, new ones would
> arise, and all we could agree upon would be, that clouds

Earth and sky divided: a final flash of evening sunlight illuminates the underside of a low-lying bank of stratocumulus, the most widespread cloud formation on earth.

there were, but that I was grossly mistaken in the *zoography* and *topography* of them.[3]

Clouds, by this reckoning, are beyond discussion, let alone any systematic attempt at representation. But, as this book has already shown, it is the inconstancy of clouds that has rendered them objects of sustained (if sometimes frustrated) attention from the many scientists, writers, artists and musicians who have responded to their nebulous challenge. So far this book has dwelt mainly on the scientific and linguistic representations of clouds, looking at the ways in which they have been named and understood, but this chapter – and much of the next – will explore some of the visual and auditory responses to clouds that have been made by artists, film-makers and musicians over the centuries.

Clouds in the visual arts

In October 1961 an exhibition entitled 'The Last Works of Henri Matisse' opened at New York's Museum of Modern Art (MOMA). By early December, nearly 120,000 people had visited it, including the late artist's son, Pierre, but nobody had yet noticed that one of the exhibits, *Le Bateau* (1953), had been hung upside down. The work, arranged from cut-out pieces of painted paper, depicts a cloud and a blue-sailed yacht reflected in the calm waters of a lake, and it was evidently the cloudy reflection that had confused

Skyscrapers act as high-rise mirrors, their tinted windows bringing the theatre of clouds down to street level.

the museum's curators. It was not until a visitor noticed the error and reported it to a museum guard that the exhibit was rehung in its correct orientation, with the more complex cloud motif restored to the top of the image. As MOMA's curators pointed out in their defence, the position of the labels and screw holes on the frame suggested that the cut-out had been hung incorrectly for most of its life, but they agreed that they should have realized that Matisse would never have rendered a reflected motif in a more complex manner than its 'original'.

The episode can be read as epitomizing the central argument of Hubert Damisch's influential study *A Theory of /Cloud/* (1972), which contends that, at a certain point in their development, images of clouds in European painting (including architectural ornamentation and set design) broke with representational convention and instead became images of unrepresentability. The clouds that had served as portals to the divine realm in medieval imagery began to play a more ambiguous role during the Renaissance, when the rise of linear perspective introduced new representational demands, to which clouds – with their formless combination of the ephemeral and the material – seemed visually unsuited. Clouds, according to Damisch, became images without likeness, visual indications of something that cannot be depicted, hence the typographical quirk of '/Cloud/'. As Mary Jacobus observes, 'By the use of two forward slashes, Damisch transforms /cloud/ into an index or signifier, rather than a word denoting "cloud" in any descriptive or figurative sense.'[4]

Damisch takes the clouds of Correggio as a case in point, noting how the 'solid-looking' clouds that the artist frescoed inside the dome of Parma Cathedral 'obscure most of the splendor and light that reign beyond them'. Their visual function is to close behind the ascending Virgin, to complete the picture (and her assumption into celestial glory) by hiding her from view.[5] Correggio's clouds mediate between the viewer on earth and the unseeable prospect of heaven by negating the architectural space in which they appear.

As Damisch observed, cloud perspective presented a particular set of challenges to visual artists. Leonardo da Vinci, in his posthumously compiled *Treatise on Painting* (*c.* 1540), offered detailed advice on depicting clouds according to the new rules of painterly perspective:

> Of two objects at equal height, that which is the farthest off will appear the lowest . . . This happens also when, on account of the rays of the setting or rising sun a dark cloud appears higher than another which is light.[6]

Yet Leonardo's perspectival cloud figures were conspicuously generic, their cotton-wool shapes more or less unrelated to the complex forms that appear in clouded skies. The pictorial evolution of clouds in Western art has been outlined many times, most recently by John E. Thornes, who argued that most clouds painted before the seventeenth century ('when the sky began to inspire artists for its own sake') were decorative entities with few formal or structural relations to the images they happened to appear in.[7] As the critic John Ruskin had complained in the first instalment of *Modern Painters* (1843),

> with them [the old masters], cloud is cloud, and blue is blue, and no kind of connection between them is ever hinted at. The sky is thought of as a clear, high material dome, the clouds as separate bodies suspended beneath it.[8]

It was Dutch and Flemish landscape painters who first brought skies fully into the frame, with images such as Jacob van Ruisdael's *Tempest,* or his *Vessels in a Fresh Breeze* (both *c.* 1660s), dominated by integrated cloud compositions that routinely covered more than two-thirds of the canvas. These artists took their skies unusually seriously and, a century before John Constable followed in his footsteps, the London-based marine painter Willem van de Velde made a series of *plein-air* cloud sketches on Hampstead Heath. According to Gilpin, he also hired a waterman to row him onto the Thames in all weathers: 'These expeditions he called, in his Dutch manner of speaking, going a-skoying.'[9]

Antonio da Correggio, detail of the fresco of the *Assumption of the Virgin, c.* 1526–30, in the dome of Parma Cathedral.

Constable evidently learned as much about clouds from looking at Ruisdael as he did from reading Luke Howard, and his response to one of the Dutchman's maritime scenes, as delivered in a lecture in London in 1836, became one of his best-known statements of painterly intent:

> The subject is the mouth of a Dutch river, without a single feature of grandeur in the scenery; but the stormy sky, the grouping of the vessels, and the breaking of the sea, make the picture one of the most impressive ever painted.
> 'It is the Soul that sees; the outward eyes
> Present the object, but the Mind descries,'
> *We see nothing truly till we understand it.*[10]

Constable's conviction, that apprehension depends on understanding, guided his meteorological adventures on Hampstead Heath, where, during the summers of 1821 and 1822, he produced more than a hundred oil sketches of clouds and skies, which are now among his most revered works. Each summer morning, he

A group of interceding angels borne aloft by a cloud in Giovanni Battista Tiepolo's *St Thecla Praying for the Plague-stricken*, late 1750s, oil on canvas. This painting commemorates the Italian plague of 1629–31.

walked the short distance from his lodgings to the meadows of Prospect Walk, an elevated spot that was ideal for cloud watching. By returning to the same locations at the same times of day, Constable was able to build up a calibrated picture of the sky over time, a fieldwork methodology drawn from the natural sciences. As Constable tellingly observed, 'painting is a science and should be pursued as an inquiry into the laws of nature. Why, then, may not landscape painting be considered a branch of natural philosophy, of which pictures are but the experiments?'[11] The diligent weather notes that Constable wrote on the back of each sketch confirm that his interest in clouds was more than simply visual, as does the presence among his personal effects of a copy of the second edition of Thomas Forster's *Researches About Atmospheric Phænomena* (1815), the first chapter of which gives a summary of Luke Howard's classification of clouds. As John E. Thornes has shown, detailed annotations to the cloud chapter show that 'Constable had a remarkable knowledge of contemporary meteorological science', which he acquired in the pursuit of

Jacob van Ruisdael, *A Tempest*, c. 1660s, oil on canvas. Ruisdael's skies were particularly admired by Constable.

accuracy and understanding in his representation of skies.[12] The art historian Kurt Badt described Constable's clouds as 'meteorologically more accurate than – we may say – all other clouds painted before his time', although such accuracy was not always a welcome quality. Henry Fuseli, painter and drawing tutor at the Royal Academy, complained that Constable's paintings made him want to call for his overcoat and umbrella, a sentiment later revived by Ruskin, the great defender of J.M.W. Turner, who dismissed Constable's cloud-haunted skies as 'great-coat weather, and nothing more'.[13]

Aelbert Cuyp, *Young Herdsmen with Cows*, late 1650s, oil on canvas. Grey, wind-formed clouds offer a visual rhyme with a group of cattle in this intriguing landscape study.

In addition to his *plein-air* cloud studies, Constable made copies of other artists' skies, including Alexander Cozens's cloud sequences that appeared in his influential textbook, *A New Method of Assisting the Invention in Drawing Original*

Compositions of Landscape (1785). Cozens had devised a kind of visual classification of cloudiness, a twenty-part sequence that ran from 'All plain' to 'All cloudy', via 'Streaky Clouds at the top of the Sky', 'Streaky Clouds at the bottom of the sky' and 'Half Cloud half plain'. Constable copied all twenty of Cozens's sky stages, complete with their long inscriptions, further evidence of his extraordinary commitment to the visual understanding of clouds. When Ruskin suggested, in 1844, that 'every class of rock, earth, and cloud must be known by the painter, with geologic and meteorologic accuracy', he was, ironically, describing the research-led practice that had been pioneered by Constable a generation before.[14]

John Constable, *Study of Cirrus Clouds*, 1822, oil on paper. One of more than 100 rapidly executed oil sketches made by Constable on Hampstead Heath during the summers of 1821 and 1822.

Like Constable, Ruskin was a cloud obsessive. He joined the London (later, Royal) Meteorological Society at the age of seventeen, and one of his first publications, 'Remarks on the Present State of Meteorological Science', appeared in the first volume of the Society's *Transactions* in 1839. He went on to write long and enthusiastic paeans to 'the Truth of Skies', by which he meant a kind of rhapsodic spiritual covenant between mankind,

How to draw clouds: '35. All cloudy, except a narrow opening at the top of the sky' and '36. The same as the last, but darker at the bottom' from Alexander Cozens, *A New Method of Assisting the Invention in Drawing Original Compositions of Landscape* (*c.* 1785).

nature and the divine. Clouds, for Ruskin, nourished both the intellect and the soul, and several chapters of his multi-volume study *Modern Painters* (1843–60) were devoted to discussions of 'The Region of the Cirrus', 'The Central Cloud-Region', or 'The Cloud-Chariots . . . the true cumulus, the most majestic of all clouds, and almost the only one which attracts the notice of ordinary observers'.[15] While Ruskin was familiar with the tenfold cloud classification, he maintained (as did Ralph Abercromby) that there were only two basic families of clouds, 'massive' and 'striated', though he conceded that some clouds could be thought of as 'fleecy':

> The fleece may be so bright as to look like flying thistle-down, or so diffused as to show no visible outline at all. Still, if it is all of one common texture, like a handful of wool, or a wreath of smoke, I call it massive. On the other hand, if divided by parallel lines, so as to look more or less like spun-glass, I call it striated.[16]

John Ruskin, *Study of Dawn: Purple Clouds*, 1868, watercolour sketch. Ruskin, who joined the London Meteorological Society at the age of 17, argued that 'every class of rock, earth, and cloud must be known by the painter, with geologic and meteorologic accuracy'.

Ruskin considered clouds to be aesthetic as much as natural phenomena, whose ultimate purpose was to supply earthbound humanity with spiritual and emotional nourishment. His own role as a critic was thus 'to demonstrate the wonder of the sky to people who seemed not to notice it', to raise their eyes heaven-wards in emulation of the wise innocence of medieval crafts-people, who 'never painted a cloud but with the purpose of placing an angel upon it'. To contrast that with the materialism

of today, he wrote, is to share his regret that 'we have no belief that the clouds contain more than so many inches of rain or hail'.[17] In one remarkable passage, written some time in the late 1850s, Ruskin recalled a winter morning spent studying a sky covered with high cirrocumulus cloudlets. As 'each was more than usually distinct in separation from its neighbour', they were easily counted, which Ruskin duly did – all 50,000 of them, likening them in his cloud-counting rhapsody to 'Flocks of Admetus under Apollo's keeping':

> Who else could shepherd such? He by day, dog Sirius by night; or huntress Diana herself – her bright arrows driving away the clouds of prey that would ravage her fair flocks. We must leave fancies, however; these wonderful clouds need close looking at. I will try to draw one or two of them before they fade.[18]

But not all clouds were as benign as those herds of winter cirrocumulus. By the time he came to write *The Storm-cloud of the Nineteenth Century* in 1884, the ageing Ruskin was prey to visions of a polluting miasma brought on by the modern, industrialized world's spiritual and environmental decline. There were 'diabolic clouds over everything', he wrote:

> In healthy weather, the sun is hidden behind a cloud, as it is behind a tree; and, when the cloud is past, it comes out again, as bright as before. But in plague-wind, the sun is choked out of the whole heaven, all day long, by a cloud which may be a thousand miles square and five miles deep.[19]

As a youth, Ruskin had declared that there was no such thing as bad weather, but his temperament had been altered by time and torment, and now he was literally under a cloud – 'Not [a] rain-cloud', he wrote, 'but a dry black veil which no ray of sunshine can pierce.'[20] Ruskin's oppressive industrial smog, his account of which has been widely interpreted as an early environmentalist manifesto, makes a better-known appearance in Virginia Woolf's time-travelling novel *Orlando* (1928), in which the turn of the eighteenth century is marked by the appearance of Ruskin's obliterating welter of cloud:

> Orlando then for the first time noticed a small cloud gathered behind the dome of St Paul's. As the strokes sounded, the cloud increased, and she saw it darken and spread with extraordinary speed. At the same time a light breeze rose and by the time the sixth stroke of midnight had struck the whole of the eastern sky was covered with an irregular moving darkness, though the sky to the west and north stayed clear as ever. Then the cloud spread north. Height upon height above the city was engulfed by it . . . As the ninth, tenth, and eleventh strokes struck, a huge blackness sprawled over the whole of London. With the twelfth stroke of midnight, the darkness was complete. A turbulent welter of cloud covered the city. All was darkness; all was doubt; all was confusion. The Eighteenth century was over; the Nineteenth century had begun.[21]

If Constable was the nineteenth century's most cloud-haunted painter, René Magritte was his twentieth-century counterpart. Clouds, along with pipes and bowler hats, were one of the Belgian surrealist's recurring motifs, often appearing in regular patterns, as in *The Curse* (1931), the first of his series of 'pure' cloud studies, or as single looming manifestations, as in *The Great Adventure* (1938), in which a cloud silently lets itself into a room, the outside invading the inside.[22] *Threatening Weather* (1929) was a surrealist assemblage in vaporous form, featuring a cloud chair, a cloud tuba and a headless cloud statue floating whitely above the sea, a precedent reversed in *The Future of Statues* (*c.* 1937), a plaster reproduction of Napoleon's death mask, over which painted clouds pass like the implacable weather of history. Many other artists, such as Emil Nolde, developed a modernist interest in clouds and skies, but Magritte made them his personal trademark, returning again and again to the paradox of clouds, free-floating objects with volume but no surface, provisional,

Jasper Cropsey, *Passing Shower on the Hudson*, 1885, oil on canvas. Cropsey's rhapsodic essay 'Up Among the Clouds' (1855) urged his fellow American artists to pay more attention to the 'dream-world' of the skies.

unfathomable and mysterious. His work served to epitomize a telling complaint made by the Victorian meteorologist William Clement Ley that painters, unlike photographers, 'seem addicted, as a class (they must forgive me for saying it), to the habit of representing the impossible in their cloud portraits'.[23] Though insightful, this was not intended as a compliment.

Clouds in photography

The advent of photography in the mid-nineteenth century brought a new documentary focus to the sky, although practical difficulties were encountered from the outset. Clouds might seem naturally photogenic, but they presented profound technical challenges to early photographers whose minutes-long exposures led to landscape studies being marred by overexposed skies. The main difficulty lay in overcoming the lack of contrast between the blues of the sky and the whites (or greys) of the clouds, which were virtually indistinguishable on early photographic plates. Dark, dramatic cumulus clouds were not such a problem, but lighter forms of cloud such as cirrus or cirrocumulus proved especially difficult to photograph. One solution, to which many early practitioners resorted, was to take separate, underexposed prints of the sky and then to superimpose them over the whited-out originals. Roger Fenton's *September Clouds* (1859) has become a much-exhibited image in its own right, but it started life as one of many expedient skies, shot to supply his studio with skyscapes to order.

Later techniques involved the use of coloured filters to counteract the dominance of daylight. As the French meteorologist and photographer Alfred Angot explained in an article in the journal *Nature* in 1896, yellow filters produced the best results, since

René Magritte,
The Future of Statues,
c. 1937, a painted
plaster reproduction
of Napoleon's death
mask.

the light coming from the sky contains very few yellow and green rays, and is thus extinguished to a great extent; but, on the other hand, the great proportion of yellow and green rays which exists in the white light of the clouds passes the screen and makes an impression on the plate.[24]

The Swiss meteorologist Albert Riggenbach had successfully experimented with the use of a Nicol prism to filter polarized light, thereby increasing the contrast of even the faintest clouds, and his black and white photographs remain some of the clearest images of clouds ever made. It was also found that developing fluids with a high proportion of bromide gave 'a greater contrast between the clouds and the sky, and the development can be carried further without fear of fogging'.[25]

Roger Fenton, *Landscape with Clouds*, c. 1856. One of a series of stock images of skies that Fenton used in his landscape work.

But not all meteorologists were happy with the results of such ingenuity. Sir Norman Lockyer, himself a pioneer of cloud photography who had worked for many years on an unfinished book with the promising title *In Thunderstorms with a Camera*, complained of certain photographers who resorted to using 'clouds of their own manufacture':

we have heard in the past, we know not with what truth, of artfully-placed pieces of cotton-wool on the printing frame, and of other devices, which, by judicious handling, have been

made to give an appearance remotely resembling that
of natural clouds, and that competent judges have been
deceived by these means.[26]

For Lockyer, cloud photography was a serious scientific matter,
and had nothing to do with creating 'pretty pictures, or even
accurate pictures'.[27] If photography was to be of value to meteorology, it would be as a problem-solving technology rather than
a purely representational one. And so it was beginning to turn
out, for photography had provided a solution to the enduring
problem of determining a cloud's precise altitude, by use of 'the
photo-nephograph or cloud camera', as the superintendent of
the Kew Observatory described it.[28] If two simultaneous photographs of the same cloud were taken at two separate locations,
the height of the cloud could then be determined using simple
trigonometry. From the 1880s onwards, members of the Royal
Meteorological Society successfully employed this technique in
locations across the country, aided by the cloud camera as well
as by that other recent innovation, the telephone:

> Two observers, a suitable distance apart, and in connection
> with each other by telephone, select a cloud by arrangement
> to which each points a camera, and the simultaneous exposure
> is effected by one of the operators releasing the shutters of
> both cameras at the same instant.[29]

Lockyer was further persuaded of the benefits of photography when images of unusual wave-clouds (or '*Wogen wolken*')
began to be circulated in the 1880s. These rare clouds, which
would later be renamed Kelvin-Helmholtz billow clouds, provided visual confirmation of German physicist Hermann von
Helmholtz's recent theories of atmospheric instability, which
proposed that varyingly dense atmospheric layers move at varying speeds and directions. Photographs clearly showed how the
upper layers of these cloud formations had been shaped by wind
shear into distinctive wave-like structures, thereby proving
Helmholtz's theories, in Lockyer's estimation.

But however valuable photography was proving to the scientific understanding of clouds and weather, a host of technical and perceptual complications remained. The lack of colour resolution was a particular hindrance: as Ralph Abercromby noted in 1887, 'Though photographs are infinitely better than drawings or engravings, it must be borne in mind that both colour and distance are lost in photography, though the form is accurately represented.'[30] A lack of chiaroscuro as well as an absence of texture, volume and depth was also a recurring problem, even among images taken at sunrise or sunset, when reduced light contrast made cloud photographs relatively easy to take. As one meteorologist ruefully observed, 'The successful photographing of clouds is so entirely different from the ordinary run of photographic work that very few photographers succeed in producing even passing results.'[31] The tireless Abercromby was himself a dedicated cloud photographer – he had taken a number of cameras with him on his round-the-world cloud voyages – and wrote at length about the challenges he had faced:

The practical difficulties of cloud photography are numerous, for besides the difficulty of getting clear pictures on gelatine plates, the time required to obtain good results is very great. Sometimes half-a-dozen plates will be fogged before a good picture can be obtained, and I have often had to wait watching a cloud for a couple of hours before a suitable light could be found. Cloud forms are so transient, that sometimes after a good picture has been focussed on the ground glass, the character will have disappeared before the dark slide can be arranged.[32]

A selection of Abercromby's cloud photographs was exhibited at the Royal Meteorological Society in 1887, with one attendee describing the prints as 'the most beautiful cloud pictures he had ever seen', so it must have come as a disappointment that none of his photographs were chosen for inclusion in either of the early cloud atlases.[33]

The choice of photographs had been an editorial priority from the inception of the first cloud atlas of 1890, but the process of selecting suitable images proved the most complex of the entire undertaking. As the editors observed in their introduction,

> the sketches of cloud forms which exist are not sufficiently accurate to serve as good guides. Good photographs give more assistance; but with them there is no colour. In order that a cloud picture may be intelligible to non-specialists, the clouds and blue sky must be at least plainly distinguishable from each other.[34]

Since most cloud photographs still failed that basic test, their solution was to print full-page chromolithographs of ten specially commissioned oil paintings ('9. Cumulonimbus' is particularly luminous) alongside twelve small-format heliotypes 'from instantaneous photographs', as a way of comparing the evidential value of different forms of visual representation. The printed photographs lacked clarity and contrast, and were at best 'intermediate between pictures and diagrams', as the editors conceded:

> It must be further remarked that the pictures were taken by different processes and therefore make different impressions. One part of the pictures shows a relatively clear sky and is mostly of old date, another was obtained by means of Eosine plates, and the almost complete extinction of the blue sky by a solution of gamboge and quinine, by which the lightest clouds come out with great sharpness, such as seems almost unnatural to the human eye.[35]

The lessons learned during the production of that first, speculative, cloud atlas were not lost on the editors of the first official *International Cloud Atlas*. The Cloud Commission of the International Meteorological Committee, as the editorial board was known, put out a call for cloud photographs, and a wealth of submissions was received from around the world,

CUMULUS

Fig. 20.

Photochromotype
of cumulus clouds
from the *International
Cloud Atlas* (1896);
original photograph
taken at the Potsdam
Observatory,
31 August 1893.

with numbers boosted by the associated publicity surrounding
the first International Year of Clouds (1896). From the more
than three hundred submitted photographs the panel selected
25 prints, fifteen of which were reproduced as colour photochro-
motypes, alongside a selection of chromolithographs of illustrative oil paintings that had been commissioned from artists
working under the direct supervision of the committee. At the
time, colour photography was a new and unfamiliar technology,
and the paintings were included partly as a reassuring visual
reference for a readership unused to encountering printed – as
distinct from hand-tinted – colour photographs on the page.
It was not until later in the twentieth century that the cloud
atlases began to be illustrated solely with photographs, although
paintings and drawings (notably of birds and animals) are still a
fixture of natural history field guides today.

In 1922 the American photographer Alfred Stieglitz made
the first of the more than 220 cloud studies that would later be
known as *Equivalents*. He had attempted to photograph skies

before, without success, but a new panchromatic emulsion that had recently become available allowed a greater range of atmospheric shades and contrasts to be captured. His first sky series, *Music: A Sequence of Ten Cloud Photographs* (1922), was made with the composer Ernest Bloch in mind. Stieglitz recalled how he wanted Bloch to see the pictures and exclaim:

> Music! Music! Man, why that is music! How did you ever do that? And he would point to violins and flutes, and oboes, and brass, full of enthusiasm, and would say he'd have to write a symphony called 'Clouds.' Not like Debussy's but much, much more.[36]

When Bloch saw the series in a New York gallery later that year, his response was apparently just as Stieglitz had foreseen.

Almost all the images in the *Music* series, as well as in the rest of the *Equivalents*, were printed in such a way that the sky appears black or nearly black, creating a striking tonal contrast between it and the clouds. Some include the sun as a compositional or lighting element, but most lack any kind of visual reference point. This approach is often claimed as heralding the dawn of abstract photography, although it is worth recalling an observation made thirty years before, in the preface to the first cloud atlas, that cloud photography 'has always something unusual about it, because the attention is drawn away from the landscape. In the present case, however, the matter is reversed. In our pictures the only object of the landscape is for orientation for the sky.'[37] Perhaps all cloud photography is, by its very nature, an unavoidably abstract visual form.

Back in 1896 Alfred Angot had issued an appeal to amateur photographers to try to photograph 'clouds which strike them as having interesting shapes, noting with care the hour when they were taken, and also the direction in which the clouds appeared'.[38] A century later, his wish found its ideal outlet in the form of the Cloud Appreciation Society, an amateur organization founded in 2004 by the British author and cloudspotter Gavin Pretor-Pinney. The Society, whose manifesto pledges to 'fight the banality of "blue-sky thinking"', maintains a vast and growing online gallery of cloud photographs, many of which satisfy Angot's requirement of having 'interesting shapes'. In the foreword to the first published collection of images from the Society's archive, Pretor-Pinney noted that 'the formations we enjoy receiving most are the ones that look like things':

'a cloud in the shape of a dog barking' or 'a flying-saucer cloud'. They are perhaps the rarest of the lot. Not only did someone have to be looking up at just the right moment, and happen to have a camera with them at the time, they also needed to be in the particular frame of mind required to be able to see shapes in the clouds.[39]

The art of seeing shapes in shapeless forms is known to psychologists as *pareidolia*, and is a long-established cultural trope, from Petrarch seeing his beloved Laura's face in the clouds (in sonnet 129 of the *Canzoniere*, c. 1340), to the much-quoted *Peanuts* cartoon in which Linus and Charlie Brown are lying on their backs, gazing up at the passing clouds. When Charlie asks Linus if he can see any shapes in them, Linus replies that he has just spotted the outline of British Honduras, the profile of the American painter Thomas Eakins and a remarkably detailed tableau of the stoning of St Stephen: 'There's the Apostle Paul standing to one side. What about you, Charlie Brown?' 'I was going to say I saw a ducky and a horsie, but I changed my mind.'[40] The exchange, which was adapted into the opening scene of the feature-length animation *A Boy Named Charlie Brown* (dir. Bill Melendez, 1969), reveals an essential truth about each character via their unconscious projections onto the shifting canvas of the clouds. 'There is an universal tendency among mankind to conceive all beings like themselves', as the philosopher David Hume observed in 1757, which is why 'we find human faces in the moon, armies in the clouds, [and] ascribe malice and good will to everything that hurts or pleases us.'[41] We are what we see in the formless shapes, an idea that is developed in Coleridge's sonnet, 'Fancy in

Alfred Stieglitz, *Equivalent*, c. 1929–30, presumably gelatin silver print.

A bear getting out of bed: the art of seeing shapes in clouds is known to psychologists as *pareidolia* (from the Greek: 'alongside image').

Nubibus, or, The Poet in the Clouds', his joyful paean to cloud-gazing, composed during a seaside holiday in 1817:

O, it is pleasant, with a heart at ease,
Just after sunset, or by moonlight skies,
To make the shifting clouds be what you please,
Or let the easily persuaded eyes
Own each quaint likeness issuing from the mould
Of a friend's fancy; or with head bent low,
And cheek aslant, see rivers flow of gold
'Twixt crimson banks; and then, a traveller, go
From mount to mount, through CLOUDLAND, gorgeous land!
. . .[42]

For Coleridge, whose obsession with clouds and weather saw him piling up the hillsides in the pouring rain, pausing to bless the bounty of 'his Supreme Majesty's servants: Clouds, Waters, Sun, Moon, Stars, etc.', the smallest hint of a cloud-shape was evidently enough to send him into pareidolian rhapsodies.[43] But cloud-spotting is not for everyone: Aristophanes mocked the habit of seeing shapes in clouds – 'you've seen clouds, haven't you, shaped like centaurs and leopards and lions and such like?' – while the episode from *Peanuts* has attracted numerous pastiches over the years, including a sequence from Bill Watterson's comic strip, *Calvin and Hobbes*, in which Calvin describes his cloud as looking like 'a bunch of suspended water and ice particles', neatly reversing the direction of the analogy while encapsulating Calvin's sceptical outlook.[44] For the Cloud Appreciation Society, however, there is no such thing as a resemblance too far, and a second published collection of *Clouds That Look Like Things* (2012) featured an array of atmospheric simulacra, from the usual line-up of dogs and dolphins to more unsettling phantasms, such as Hamlet toying with Yorick's skull, a vast cloudy hand descending from the heavens and a gigantic, sunset-tinted vision of Mick Jagger's disembodied lips ('Get Off of My Cloud'), that was also a life-imitating-art recreation of Man Ray's atmospheric painting *The Lovers* (1936). As Pretor-Pinney observes,

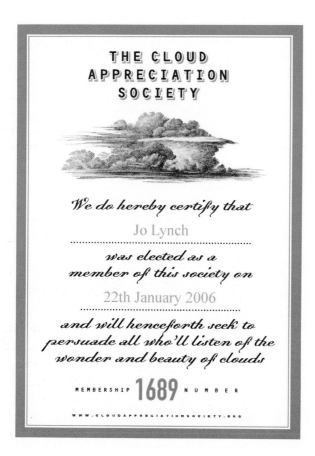

'No other organisation could have come up with a set of images like these; [an] international group of people who live with their heads in the clouds and their cameras in their pockets.'[45] It's a group of which I'm delighted to be a card-carrying member.

Tornadoes on film

In the summer of 1926, the film director (and ex-flying ace) William A. Wellman began work on his first dogfight movie, *Wings*, vowing from the outset that not a frame of his picture would either be faked in a studio or spliced with bought-in footage. All shots of aerial combat, using borrowed U.S. Air Force planes, were to be filmed as live action sequences using cameras mounted on the aircraft. But there was a problem: the skies above San Antonio, Texas, where filming was about to begin, had remained cloudless for weeks, holding up the shoot at enormous expense. After a month had passed, Paramount Pictures despatched an executive to Texas, who demanded to know why Wellman was waiting for clouds. The director's eloquent reply has gone down in Hollywood legend:

> Say you can't shoot a dogfight without clouds to a guy who doesn't know anything about flying and he thinks you're nuts. He'll say 'Why can't you?' It's unattractive. Number two, you get no sense of speed, because there's nothing there that's parallel. You need something solid behind the planes. The clouds give you that, but against a blue sky, it's like a lot of goddam flies. And photographically, it's terrible.[46]

Under pressure from the studio, Wellman's cameraman Harry Perry attempted to rig up some model clouds using cotton balls suspended on fine thread; the effect, predictably, was risible. A team of skywriters was then sent up in an attempt to create some short-lived imitation clouds, but this, too, proved a failure. Eventually, a fleet of real cumulus clouds put in an appearance, and *Wings* was completed to Wellman's satisfaction, albeit vastly over budget. The film, however, went on to win the first Best Picture Oscar in 1928, as well as the Oscar for Best Effects – vindication of Wellman's insistence on using natural clouds as a source of aerial perspective. The combat sequences remain among the best regarded in cinema history.

In the decades since *Wings* was made, film technology has changed beyond recognition, but the challenges of cloud photography have not gone away. Skies without reference points remain notoriously hard to capture on film, while the absence of clouds can mean the absence of atmosphere, as discovered by the editing team of the popular bbc television drama *Poldark* (screened in 2015). Post-production on the first series saw layers of suitably 'brooding' clouds added to a number of coastal scenes that had

been filmed during an unexpectedly cloudless Cornish summer. Similarly, the elusive cloud phenomenon known as the 'Maloja Snake', which winds its way through the Maloja Pass in the Swiss Alps, is best seen in early mornings in autumn, when warm air rises up the valley slopes and is transformed into a low, sinuous cloud formation. The cloud ribbon makes a dramatic appearance towards the end of the award-winning film *Clouds of Sils Maria* (dir. Olivier Assayas, 2014), its uncanny presence summoned partly by nature, but mostly by the visual effects designer Thomas Zolliker, who has developed a reputation as a fog and cloud specialist among European film directors.

To point a camera at a near-stationary cumulus cloud is one thing; to shoot a deadly, fast-moving tornadic stormcell is quite another. Tornadoes are violent, rotating columns of air that connect the base of a cumulonimbus storm cloud to the ground below. They form when powerful downdraughts drag the

First World War battle scene from the Oscar-winning dogfight movie *Wings* (dir. William A. Wellman, 1927), depicting the Battle of Saint-Mihiel.

A river of low-lying stratus cloud snakes its way along the Dordogne valley as dawn begins to break.

storm's revolving core towards the ground, generating a visible condensation funnel that throws up soil and debris as soon as it makes contact with the surface. Although tornadoes are common to every continent on earth, they are synonymous with the American Midwest, where they are responsible for around a hundred deaths per year. 'The populous region of the United States is forever doomed to the devastation of the tornado', observed the writer and engineer John Park Finley in 1887. 'As certain as that night follows day is the coming of the funnel-shaped cloud.'[47] Finley's landmark book *Tornadoes* (1887) was the first to be devoted to the subject, and though it was intended to be read as a manual on how to survive tornadoes, it is the remarkable illustrations to which most attention has been paid. Many of the engravings were taken from photographs supplied by the Franklin Photo-Electric Company, showing scenes of devastation following severe tornado attacks, while others were

engraved from photographs taken in the heart of the storm, or from sketches made soon after the events, showing the development of a tornado from its first tentative descent from the base of a cloud to its deadly and destructive advance across large tracts of land.

It was the advent of photography that made tornadoes familiar around the world. As Finley observed, a photograph of a tornado in action had far greater impact than a written account, especially as the descriptions tended to be contradictory: 'The tornado-cloud has been called: "balloon-shaped"; "basket-shaped"; "egg-shaped"; "trailing on the ground like an enormous kite"; of "bulbous form"; "like an elephant's trunk", etc., etc.'[48] But in spite of the technical challenges they faced, early photographers captured some impressive tornado pictures, the earliest-known example of which, taken on 26 April 1884, shows a well-defined tornado in the early stages of 'roping out' (dissipating) as it passes over the small town of Garnett, Kansas. According to a contemporary press report, 'The tornado was plainly visible from Garnett for about 30 minutes, and moved so slowly that it was successfully photographed during its progress.'[49]

The shot was taken by one A. A. Adams, who owned a photography studio in a neighbouring town. Adams went on to

The birth of a tornado: engraved photograph from John Park Finley's pioneering book *Tornadoes* (1887).

Engraving of the first
known photograph
of a tornado, taken
on 26 April 1884,
Anderson County,
Kansas.

33

in the same degree with every appearance of the cloud,
but the lower end of it (the part nearest the earth) is in-
variably the smallest. Whatever the inclination of the
central axis of the cloud to the vertical or plumb line, the lowest
end is the narrowest and nearest the earth. As seen in different positions and stages of development by various observers,
located differently, the tornado-cloud has been called : "balloon-shaped ;" "basket-shaped ;" "egg-shaped ;" "trailing
on the ground like the tail of an enormous kite ;" of "bulbous form ;" "like an elephant's trunk," etc., etc. In the
majority of instances, however, observers describe the cloud
as appearing like an upright funnel. When the tip end of

Tornado-cloud which passed near Garnett, Kansas, at 5:30 P. M. April 26,
1884. From an instantaneous photograph,

sell hundreds of souvenir photocards of the image, as well as
placing it in periodicals including *Nature*, *Science* and the
American Meteorological Journal, but, unfortunately for him, a far
more dramatic photograph of a tornado was soon in circulation.
This second image, taken by F. N. Robinson, shows a powerful
tornado kicking up a violent debris cloud near Howard City,
South Dakota, on 28 August 1884, only four months after the

The second known tornado photograph, taken in Howard City, South Dakota, on 28 August 1884. The main funnel kicks up a dust-cloud on the ground, while a pair of secondary funnels emerge like a pair of horns on either side, giving the cloud a faintly devilish air.

Garnett storm. A pair of 'satellite tornadoes' flank the main funnel, giving the rain-darkened cloud a decidedly devilish air. Unlike the Garnett tornado, the Howard storm caused a number of fatalities, and public interest in this extraordinary, almost gothic image quickly overshadowed Adams's earlier print. John Park Finley reproduced both images in his *Tornadoes* book, along with a sketch of the second tornado made by an amateur artist, J. H. Nott of Redstone, South Dakota. Finley observed, 'These two pictures of the same storm, made 20 miles apart in adjoining counties by different persons having no knowledge of each other, are valuable confirmations of one another.'[50] As with the array of visual media in the early cloud atlases, the photograph supported the testimony of the drawing, while the drawing, in turn, lent support to the testimony of the photograph.

Although Robinson's tornado image had evidently been subjected to a degree of studio manipulation, meteorologists agree that its outline, at least, was of a genuine tornado. The much-admired 'cyclone' sequence in *The Wizard of Oz* (dir. Victor Fleming, 1939), by contrast, remains one of Hollywood's most convincing tornadoes, despite being constructed from an 11-metre muslin wind-sock suspended from a steel gantry. The scene proved to be the most difficult and expensive of the entire

Engraving of a sketch of the South Dakota tornado, 28 August 1884, from J. P. Finley's *Tornadoes*. As Finley observed, 'these two pictures of the same storm, made 20 miles apart in adjoining counties by different persons having no knowledge of each other, are valuable confirmations of one another.'

•

36

violently upward by a spirally inward and upward motion which fairly crushes and grinds into pieces buildings, trees, and whatever else falls in the line of the advancing cloud. The spirally upward motion throws the ascending débris in a circular manner outward at the top of the tornado-cloud. This débris, when beyond the central whirl of the cloud, falls to the earth, but in such a manner and so disposed as to indicate the character of the force which acted upon it.

Tornado near Redstone, Davison Co., Dakota, Aug. 28, 1884. From a sketch by J. H. Nott. See opposite page. These two pictures of the same storm, made 20 miles apart in adjoining counties by different persons having no knowledge of each other, are valuable confirmations of one another.

No. II. is called the *progressive* motion of the tornado-cloud, the motion which determines the cloud's track from one point to another. The rate of progressive velocity ranks next in order to the velocity of motion No. I., although it is at all times far below the high degree of the latter.

The rate of progress of the tornado-cloud is subject to great variability throughout the path of any one storm, although on the average tornado-clouds possess a moderately uniform velocity of progression. Some observers have indicated the movement by the following expressions: " All in an instant." "Gone in a moment." "Quicker than

film – the gantry alone cost more than $12,000 – and required all the ingenuity that Arnold Gillespie (head of MGM's special effects department) could muster. His first attempt, a 1-metre black rubber cone, had failed to 'twist' convincingly, but the muslin version moved too much, and needed to be restrained. 'We had to double-lace the cyclone with music wire so it would hold together when we spun it', recalled propmaker Jack McMaster:

I was small, so they put me inside the cyclone. The men who were lacing the wire would poke their needles inside the muslin, and I would poke them back out again. It was pretty uncomfortable when we reached the narrow part.[51]

To heighten the tornadic effect, fuller's earth and compressed air were fed into the top and bottom of the cone. 'That was to create the dust cloud, the big disturbance that comes when a tornado goes along the ground', recalled Gillespie. 'The muslin was sufficiently porous so that a little of the fuller's earth sifted through, giving a kind of blur or softness to the material. That helped to keep it from looking like an artificial hard surface.'[52] Once the tornado had been filmed, the footage was processed for use as a background for the live actors. A metre-high model of Dorothy's farmhouse was then photographed falling onto a floor painted to look like the sky. To create the moment that the tornado picks it up with Dorothy and Toto inside, the film was simply run backwards.

The iconic tornado from *The Wizard of Oz* (1939) remains one of the most convincing in Hollywood history, despite being constructed from a muslin wind-sock suspended from a metal gantry.

This tornado, filmed over Oklahoma in May 1981, was dubbed the 'Wizard of Oz tornado', due to its resemblance to the snaking tornado in the 1939 film.

In an example of nature seeming to imitate art, a tornado that appeared over Cordell, Oklahoma, on 22 May 1981 was nicknamed the 'Wizard of Oz tornado' due to its unusual shape and movements, which closely resembled the behaviour of the twister in the film. Tornadoes can remain highly dangerous at the final, 'rope' stage, especially if they are being blown around by the wind, and footage of the Cordell tornado, available on YouTube, shows how vigorous its last moments were, as it spun into a sinuous curve before dissipating. The tornado was filmed as part of Mississippi State University's 'Sound Chase' project that sought to monitor the variety of noises made by tornadic storms.

In recent decades, satellite photography has transformed our understanding of clouds as atmospheric systems, their macroscopic patterns having only become apparent since the advent of earth-orbiting satellites. As historian Paul N. Edwards observed, satellite imagery meant that 'meteorologists could literally *see*

From the Creators of
"JURASSIC PARK"

and the Director of
"SPEED"

TWISTER

The Dark Side of Nature.

WARNER BROS. and UNIVERSAL PICTURES Present

AMBLIN ENTERTAINMENT Production A JAN DE BONT Film HELEN HUNT BILL PAXTON "TWISTER" JAMI GERTZ and CARY ELWES Editor MICHAEL KAHN, A.C.E.

Production Designer JOSEPH NEMEC III Director of Photography JACK N. GREEN, A.S.C. Music by MARK MANCINA Executive Producers STEVEN SPIELBERG, WALTER PARKES, LAURIE MacDONALD and GERALD R. MOLEN

Story by MICHAEL CRICHTON & ANNE-MARIE MARTIN Produced by KATHLEEN KENNEDY, IAN BRYCE and MICHAEL CRICHTON Directed by JAN DE BONT

Twister (dir. Jan de Bont, 1996) is the story of an obsessed meteorologist, played by Helen Hunt, who invents a tornado research device, named DOROTHY in homage to the original movie tornado from *The Wizard of Oz.*

A waterspout photographed from an aircraft accompanying a North Atlantic convoy during the Second World War. Waterspouts are essentially tornadoes at sea, although the name is misleading, as they consist of atmospheric cloud condensed within the whirling funnel, not seawater picked up from the surface.

large-scale weather systems, instead of laboring to construct maps and mental images'.[53] The jet stream, for example, had been known about since the 1930s, but the ribbons of cirrus cloud that track it over hundreds of kilometres were only made apparent by photographs taken from outside the earth's atmosphere, as were the swirling patterns of Von Kármán vortices (named after the physicist who first described them). Von Kármán vortices form when fast-moving currents of air flow around an obstacle in their path; from ground level, the eddies in the cloud appear as local perturbations in the stratocumulus layer, but from space the scale of the pattern is revealed by the extensive paisley swirls that spool out for many kilometres in the wake of the obstruction.

Satellite photography has also confirmed that earth is not the only planet with clouds. Mars, our nearest neighbour, is host to icy cirrus, cirrocumulus and stratocumulus clouds at its poles, while the gas giants Jupiter and Saturn each have an inner layer of watery cumulus clouds. Venus's stratiform clouds, by contrast, are composed of sulphuric acid, while the clouds of the ice giants Uranus and Neptune are mostly formed of methane. Neptune's stormy atmosphere features a prominent patch of

Martin John Callanan, *A Planetary Order* (2009): a 3D-printed terrestrial cloud globe, showing the earth's cloud cover from one second in time (2 February 2009 at 0600 UTC precisely), using information from all six cloud-monitoring satellites overseen by NASA and the European Space Agency (ESA).

Jet stream cirrus clouds over the Red Sea, as seen from the Gemini 12 spacecraft in November 1966.

Water ice clouds over Mars,
photographed from the
Viking spacecraft in 1976.

Sulphuric acid clouds on Venus, as seen by the Pioneer Venus orbiter in 1979.

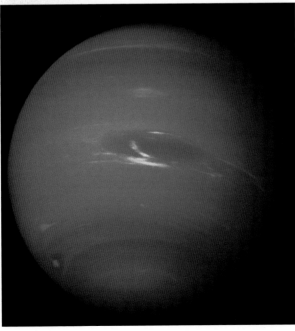

The prominent patch of cirriform clouds on the planet Neptune that NASA scientists have nicknamed 'the scooter'.

cirriform cloud – aptly nicknamed 'the scooter' – which circles the blue planet at great speed.

In August 2014 a team of astronomers detected possible signs of water-based ice clouds on a Jupiter-sized brown dwarf some 7.3 light years from earth. Of particular interest to atmospheric scientists is the fact that the planet-sized object appears only partly cloudy, like earth. 'One of the things we don't really know

This is the first image captured by Explorer vi Earth satellite, launched in August 1959. It shows a sunlit area of the Central Pacific with its cloud cover, as seen from 27,000 km (17,000 miles) above earth.

is how common partly cloudiness is', one of the astronomers commented.[54] The discovery has not yet been securely confirmed, but if it is, it will mark the first ever sighting of watery clouds beyond our own solar system.

Clouds in music

Though let us start with music in clouds. In June 1867 the French astronomer and science writer Camille Flammarion was floating in his hot-air balloon, travelling west from Paris, when the vessel entered a passage of thick, high cloud, and he could no longer tell what direction he was travelling in, or even whether he was rising or sinking:

> Suddenly, whilst we are thus suspended in the misty air, we hear an admirable concert of instrumental music, which seems to come from the cloud itself and from a distance of a few yards only from us. Our eyes endeavour to penetrate the depths of white, homogeneous, nebulous matter which surrounds us in every direction. We listen with no little astonishment to the sounds of the mysterious orchestra.[55]

As Flammarion had already noted, the cloud he was in had yielded a particularly high hygrometer reading, and it was this high humidity that had served to funnel the sounds from a band playing in a town square more than a kilometre below. Many other balloonists have described similar encounters with the surprising soundscape of the upper air. James Glaisher claimed to have heard 'a band of music' playing at an elevation of nearly 4 km (13,000 ft) during his first balloon ascent, over Wolverhampton, in July 1862.[56]

The acoustic resonance of the upper atmosphere is something to which sound artists have only recently begun to respond. In the summer of 2004 the British architect Usman Haque launched a remarkable sound installation named *Sky Ear*, a cloud of 1,000 helium balloons fitted with a payload of mobile phones, sensor circuits and flashing LEDs. The phones

A cumulus cloud stretches out over Sydney Harbour, as seen from a descending airliner. Although it appears to be lighter than air, a cloud of this size contains as much as 200 tonnes of water.

were set to auto-answer, allowing observers to call them up and listen to the sounds of the troposphere, literalizing Mark Strand's poetic conceit that 'one speaks into a cloud as one would a telephone.'[57] What emerged was an eerie symphony of whistles and hums picked up from the electromagnetic Babel that fills the sky like an artificial atmosphere, your call from the ground supplying a momentary connection to the stream of human-generated hertzian weather that passes through the skies overhead.

In 1997 a Canadian architect named Nicholas Reeves designed a 'meteo-electronic instrument', or cloud harp, which works in a similar way to a CD player. A skyward-pointing infrared laser 'reads' the clouds as they pass above the installation, converting their patterns into audio and musical sequences, the varying tones of which correspond to each cloud's altitude and density.

Musicians and songwriters have long looked to clouds for content as well as inspiration. A territorial (if ungrammatical) Mick Jagger told the world to get off of his cloud in 1965 while, as was seen at the end of Chapter One, Joni Mitchell saw clouds from both sides, in the standout track from her second album, *Clouds* (1969), with its melancholy admission that she really did not understand clouds at all. An intriguing nephological connection, meanwhile, links the British punk band The Stranglers with Hugo H. Hildebrandsson, one of the co-authors of the first *International Cloud Atlas*. 'Sweden', a track from the Stranglers' third studio album, *Black and White* (1978), contains a reference to distant cumulonimbus formations, along with the claim that Sweden is the only country with interesting clouds. The lyrics were written by vocalist Hugh Cornwell, who had spent two years studying for a PhD in biochemistry at the University of Lund, only a few hours from Uppsala, where Hildebrandsson had been professor of meteorology. Ralph Abercromby had travelled to the University of Uppsala in 1887 in order to work with Hildebrandsson on their revised cloud classification, so the word 'cumulonimbus' actually entered the global language of clouds in Sweden, a country where (according to Cornwell) 'there's so much time to muse about life you start concentrating on cloud formations'.[58] Sadly, this was not intended as a compliment.

A few years later, ambient pioneers The Orb sampled an interview with the American vocalist Rickie Lee Jones for their iconic track 'Little Fluffy Clouds' (1990), in which the singer recalled the fiery, multi-coloured skies of her Arizona childhood in lilting, Hopkinsesque detail. A 'Cumulonimbus Remix' of the track was released in 1991, its form and content closely reminiscent of the extended 'Meteorological Mix' of Kate Bush's 1986 single 'The Big Sky', which sampled a series of vox-pop comments likening passing cumulus clouds to trees, industrial waste or a map of Ireland (an image that recalls William Gladstone's notorious description of Ireland as 'that cloud in the west, that coming storm'). As Bush later recalled, the song was written during a sojourn in rural Ireland, where she would 'watch the clouds rolling up the hills towards me; there's a lot of weather on this album'.[59] 'Cloudbusting', a track from the same album (*Hounds of Love*, 1985), tells the similarly meteorological story of the controversial psychoanalyst Wilhelm Reich, who designed a portable machine named 'the cloud-buster' that he claimed could produce (or avert) rain by altering local levels of an atmospheric life-energy that he termed 'orgone'. In July 1953 the blueberry farmers of Bangor, Maine, hired Reich and his mysterious machine in the hope of ending the summer drought that was threatening their crop. Reich set up his cloud-buster on the shore of West Grand Lake and pointed it at the sky for a little over an hour. The following morning it began to rain, and the crop was saved. Reich, of course, claimed victory over the elements, with a local newspaper quoting eyewitness testimony that 'the queerest looking clouds you ever saw began to form soon after they got the thing rolling'.[60] It was to be Reich's last triumph, however, for three years later he died in prison in Pennsylvania, having been convicted of 'fraud of the first magnitude' in connection with a variety of unorthodox orgone-based 'therapies' to which he had subjected his unfortunate patients.

Although clouds are (mostly) silent, their ceaseless alterations in shape and texture render them ideal subjects for orchestral compositions. Claude Debussy described how *Nuages* ('Clouds'), the first of his three *Nocturnes* (1889), sought to render 'the immutable aspect of the sky and the slow, solemn motion of the clouds'. The piece, he recalled, had been prompted by the memory of an evening walk in Paris, when 'some clouds slowly pass[ed] through a moonless sky, a number of clouds, not too heavy, not too light: some clouds, that is all'.[61] This cloud memory shaped the structure of the eight-minute piece, which starts with a slow-moving, chordal theme, followed by a second, airier theme, as introduced by a flute and harp. In an insightful review of the first performance of *Nuages*, in Paris, the composer and critic Paul Dukas noted that Debussy's 'clouds' were more metaphorical than meteorological:

In the first of these *Nocturnes*, the 'décor' consists of unfurling clouds on an unchanging sky, their slow progress achieving 'in anguished grey tones lightly tinged with white' music which does not set out to be a meteorological representation of such a phenomenon, as one might think. It is true that it makes allusion to it through the continual floating of sumptuous chords whose rising and falling progressions recall the architecture of the skies . . . However, the real significance of the piece still remains symbolic: it translates analogy through analogy in the medium of music in which all the elements, harmony, rhythm and melody, seem in some way to have vanished in the ether of the symbol, as if reduced to an imponderable state.[62]

Like Debussy, the Hungarian-born composer György Ligeti remains best known for his atmospherically charged works, notably *Atmosphères* (1961) and *Clocks and Clouds* (1972), the latter composed in response to the philosopher Karl Popper's ideas concerning the varying rhythms of nature, as outlined in his landmark essay *Of Clouds and Clocks* (1966). Popper had invoked two distinct kinds of processes occurring in nature: one that can be measured exactly ('clocks') and the other made up of indefinite occurrences that can only be approximated ('clouds').

Mara Carlyle and Lisa Knapp perform cloud-themed music at 'Escape to the Clouds', the inaugural Cloud Appreciation Society conference held in London in September 2015.

'My clouds are intended to represent physical systems which, like gases, are highly irregular, disorderly, and more or less unpredictable', wrote Popper:

> According to what I may call the commonsense view of things, some natural phenomena, such as the weather, or the comings and goings of clouds, are hard to predict: we speak of the 'vagaries of the weather'. On the other hand, we speak of 'clockwork precision' if we wish to describe a highly regular and predictable phenomenon.[63]

The ostentatious clouds of opera: the multi-canopied cloud machines from Giacomo Torelli's designs for a mid-17th-century Venetian opera – Francesco Sacrati, *La Venere Gelosa* (1643) – features Apollo's palace as a visionary city amid the clouds.

Yet the boundaries between clouds and clocks are not always so precise: the changing seasons are 'unreliable clocks' rather than clouds; plants are more clock-like than animals, while a puppy is more cloud-like than a mature, obedient dog. On the other hand, a strictly mechanical view of the universe would hold that 'all clouds are clocks – even the most cloudy of clouds', which are only unpredictable because we don't understand them.[64]

According to Ligeti's biographer, Richard Steinitz, the composer was much taken with Popper's suggestive imagery, especially the conundrum that not only are all clocks 'cloudy' (that is, imperfect to some degree), but clouds are 'clandestine clocks (only appearing to be clouds because we are ignorant about the interaction of their particles)'.[65] *Clocks and Clouds* (the title reverses Popper's) is a thirteen-minute piece for chorus and orchestra, in which the music glides fluidly between 'clockiness and cloudiness', in Steinitz's words, in a musical exploration of the elemental tensions between solid precision and misty imprecision: 'For the moment, he was attracted to the concept of clocks melting into clouds, and the reverse, clouds solidifying into clocks. Using his musical transformation techniques, he

Nicola Sabbatini's treatise on theatrical machinery reveals how the effects were achieved: large wooden arms worked by ropes and pulleys moved the rolls of painted cumulus clouds into position, creating the realm of the gods on earth.

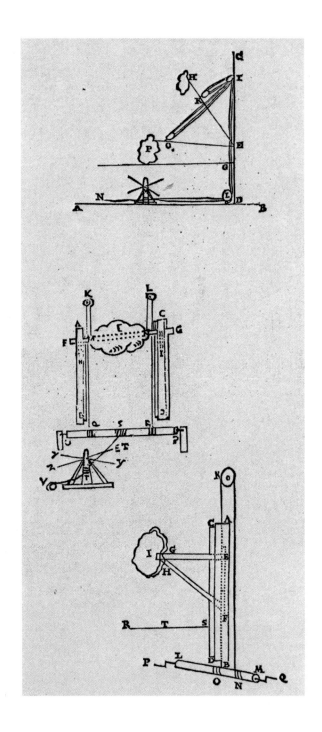

Clouds and
clocktowers:
a low bank of
stratocumulus
dumping winter
snow on the hills
behind Reykjavik,
Iceland.

for Gavin Osborn

The invention of clouds

Nina Whiteman

wanted to write the slowest and smoothest of metamorphoses.'[66] Like a cloud itself, the piece appears simultaneously simple and complex, moving and static, ordered and chaotic, with a riot of small-scale movements occurring inside a larger kinetic structure. It is a breath-taking piece of music.

And clouds continue to haunt the musical imagination. The British composer Nina Whiteman has drawn on a range of sources for her recent work, including meteorology. Her three-part cycle for bass flute, percussion and voice, *The Invention of Clouds* (2009), *The Modifications of Clouds* (2009) and *Night Shining* (2011), took initial inspiration from Luke Howard's cloud classification, the complex, allusive language of which lends shape to the sinuous sound world of the pieces. A number of Howard's phrases have been incorporated into Whiteman's scores, initially as notations of musical texture and timbre ('an irregular spot first appears . . . convex or conical heaps'), but later,

Nina Whiteman's score for *The Modifications of Clouds* (2009), incorporating phrases from Luke Howard's 1803 essay.

in *The Modifications of Clouds*, as sung words and phrases: 'Nubes cumulata, densa, sursum crescens'; 'Nubes cirrata, tenuissima, quae undique crescat', a secular incantation to guide a twenty-first-century audience through an auditory journey into cloudland.

5 Future Clouds

This book is being written in the cloud. That means that my word
processor is not installed on my computer. The files for each chapter
are not being saved on one of my own hard disks or USB keys.
Instead, the program I am using is running somewhere 'out there'
on the Internet.
Christopher Barnatt, *A Brief Guide to Cloud Computing* (2010)[1]

This book is *not* being written in the cloud, as it happens, but my
next one almost certainly will be, if the adoption curve described
in Christopher Barnatt's *Brief Guide to Cloud Computing* (2010)
is anything to go by. By 2020, he writes, 'cloud computing' will
simply be known as 'computing', and everyone – even laggards
like me – will be running their digital lives through the cloud.[2]
This transformation sounds a lot more elemental than it
really is. 'The cloud', after all, is just a network of servers, but a
certain meteorological mystique has attached itself to the term,
reinforced by images of laptops and tablets perched on little
white clouds like ancient gods or immortals. The term 'cloud',
in its computing sense, has been in use among data technicians
since the 1990s, but it became a household word in the 2010s, in
the wake of Apple co-founder Steve Jobs's much-reported
declaration that 'we are going to move the digital hub, the center
of your digital life, into the cloud', in a cybernetic reworking of
De Quincey's vision of the Brocken spectre as a nebulous
repository of secrets and desires.[3] Ten years ago any Internet
search for 'cloud' brought up only weather sites. Now it brings
up Google Cloud and Microsoft Azure, ethereal-sounding,
eco-positive names that mask the physical reality of their data
centres' vast, polluting infrastructures, whose energy use far
outstrips those of the most carbon-emitting industrial plants
on the planet. As the film-maker Timo Arnall points out, 'It's
easy to forget that the cloud is a thing run by humans, and it (a
machine) has to be stored someplace, somewhere.'[4] Arnall's 3D

documentary *Internet Machine* (2014) explores the monumental architecture of the cloud servers, revealing the materiality of a technology that trades on metaphors of immateriality: 'ether', 'cloud' and, more recently, 'fog computing' or 'fogging'. A headline in the *Wall Street Journal* in May 2014 read: 'Forget "the Cloud"; "the Fog" is Tech's Future'. But then, as this final chapter sets out to show, the limitless digital cloudmosphere is only the latest of the myriad ways in which future clouds – and future weather – have been imagined and created.

Cloud engineering

'We are going to move the digital hub, the center of your digital life, into the cloud': Apple CEO Steve Jobs's 2010 announcement did much to publicize the concept of cloud computing.

Attempts to engineer clouds and weather go back a surprisingly long way. In the 1840s the American meteorologist James Espy proposed large-scale fire-burning in order to precipitate rain from the resulting pyrocumulus clouds, while retired Civil War general Edward Powers, in his book *War and the Weather, or, The Artificial Production of Rain* (1871), recommended the detonation of shells and explosives high in the sky as a means of stimulating

the rain that he had often observed falling in the aftermath of battle. In 1880 he even secured a patent 'for producing rain fall by conveying and exploding torpedoes or other explosive agents within the cloud realm'.[5] Powers's ideas were never put into practice, although in his memoir *Cider With Rosie* (1959), Laurie Lee recalled an episode during the dry summer of 1921 when 'soldiers with rifles marched to the tops of the hills and began shooting at passing clouds'.[6] Heavy rain soon followed, ending the long summer drought, but whether it was from the gunshots or 'a simple return of nature' will never be known.

By the 1940s artificial rainmaking was being trialled in earnest by two American scientists, Irving Langmuir and Vincent Schaefer of the General Electric Company's research laboratories at Schenectady, New York. They had developed a technology they called 'cloud seeding'. This was based on their discovery that water droplets within certain clouds remain in a 'supercooled' liquid state at temperatures well below freezing, until some form of external nuclei stimulates the droplets' transition into ice. By exhaling into a specially designed open-topped freezer, Langmuir and Schaefer were able to create artificial clouds on which to perform their experiments. At first, all their 'rainmaking' attempts failed, until Schaefer added a block of dry ice (frozen CO_2) to the chamber, at which point 'his breath began to shimmer and sparkle in the light. It had instantly turned into crystals, which fell to the floor with the same dendrite shapes found in natural snow.'[7]

It was not long before similar experiments were being conducted in the upper air, with dry ice scattered from a crop-duster plane flying just above the clouds. The first trial, in November 1946, worked spectacularly, with snowflakes falling from the affected cloud within minutes of seeding, leaving a hole where the droplets had frozen and fallen away; the cloud had 'almost exploded', as Schaefer recorded in his notebook.[8] To prove that the holes would not have formed anyway, dry ice was once used to cut the General Electric logo from a bank of supercooled stratus, creating the world's first commercially branded cloud.

By 1947 funding for the experiments had been taken over by the U.S. government, under the code name Project Cirrus, with the unsettling aim of harnessing clouds and their contents for military use. By then silver iodide had replaced dry ice as the chemical of choice for large-scale cloud modification, since when many thousands of tonnes of the compound have been sprayed onto the summits of clouds around the world. The first public confirmation of these secret military trials came in July 1972, when the front page of the *New York Times* claimed that

Project Stormfury was a u.s. government-funded initiative that experimented with weather modification. This photograph was taken in September 1969, during one of the project's research flights.

American planes had for several years been covertly seeding clouds over Southeast Asia in an effort to precipitate flash floods over the Ho Chi Minh Trail, the Vietcong's vital network of jungle supply routes. While it cannot be known for certain whether the military's efforts were successful, the project's lead scientist, Pierre St Amand, recalled how 'the first cloud we seeded grew like an atomic bomb explosion, and it rained very heavily out of it and everybody was convinced with that one experiment that we'd done enough'.[9]

The world soon agreed that they had done enough, and in 1977 a multilateral convention banning weather modification for military purposes was signed by forty countries, including the United States, and today most cloud seeding is used to irrigate arid farming regions such as Western Australia, though the technology is occasionally deployed by ski resorts in an effort to induce early-season snowfall. In Beijing clouds are routinely seeded in advance of major public holidays, such as China's National Day (1 October), in order to encourage rain clouds to empty themselves in advance. The run-up to the opening ceremony of the Beijing Olympics in August 2008 (the height of northern China's rainy season) saw a barrage of more than a thousand silver iodide rockets launched at a looming cloud belt from artillery batteries mounted in the suburbs. Baoding City, southwest of Beijing, received about 100 mm of rain that night, but not a drop fell on the roofless Olympic stadium for the entirety of the Games.

Rainmaking in China: officials from central China's Hubei province shoot chemical rounds into clouds from an anti-aircraft gun, May 2011.

The opening ceremony of the London Olympics in 2012, by contrast, saw a fleet of large polystyrene cumulus clouds installed inside the stadium as part of Danny Boyle's 'Isles of Wonder' pageant, in acknowledgement of the role that rain has played in the shaping of the British character. 'They will be real clouds that will be hanging over the stadium', Boyle explained on the eve of the Games. 'We know we're an island culture and an island climate. One of these clouds will provide rain on the evening, just in case it doesn't rain.'[10] In the event, no rain fell from either the real or the artificial clouds, even during Kenneth Branagh's impressive rendition of Caliban's speech from *The Tempest* (1611), in which 'the clouds methought would open and show riches ready to drop upon me, that when I waked I cried to dream again' (III.2).

Weather modification remains a contentious area, partly for environmental reasons – there may be unforeseen side effects from spraying chemicals into clouds – and partly over questions of ownership: who has the right to harvest rain from clouds that pass over national borders? In 1949 a minor geopolitical dispute flared up between Canada and the United States, following a cloud seeding experiment near the Montana/Saskatchewan border. Canada claimed that rain-bearing clouds had been drifting towards their arid wheat prairies, but that the American rainmakers had, in effect, 'stolen' the rain for themselves. The dispute led to a UN treaty limiting the artificial creation of rain near the Canadian border, and to a flurry of articles in American legal publications with provocative titles such as 'Who Owns the Clouds?' and 'Legal Remedies for Cloud Seeding Activities: Nuisance or Trespass?'[11] In July 2004, in the wake of a regional cloud-seeding operation in central China's drought-stricken Henan province, an argument broke out between two neighbouring cities, with one accusing the other of stealing its rain. As an official from the city of Zhoukou complained, meteorologists in nearby Pingdingshan had intercepted rain clouds that were on their way to Zhoukou, emptying them with their seeding cannons. While Pingdingshan received more than 100 mm of rainfall that day, less than 30 mm fell on Zhoukou.[12] As the stresses on agricultural land increase with rising populations, competition over access to rainwater looks set to escalate, and the current legal formulation of clouds as 'a resource that belong to no one, and therefore susceptible to appropriation by anyone' may well have to be rewritten.[13]

Anthropogenic clouds

Some of the most commonly seen clouds in the skies today are anthropogenic in origin. Entire skyscapes of spread-out aircraft contrails routinely appear over busy flight paths, with satellite photography confirming the extraordinary scale and prevalence of these man-made ribbons of cloud. Contrails (a contraction of 'condensation trails') are formed from water vapour and microscopic particulates ejected by aircraft exhausts, usually at altitudes above 8 km (26,000 ft) and, like natural cirrus clouds, they are mainly composed of slowly falling ice crystals. They can persist and spread over great distances – sometimes as far as 150 km (93 miles) – depending on the prevailing winds, as well as on the amount of moisture already present in the atmosphere. If the air is very dry, short-lived contrails will briefly trail an aircraft before evaporating, but if the air is already full of vapour, contrails can quickly spread over the sky in a chain reaction, nucleating the surrounding vapour into streams of visible cloud.

Contrails have become so prevalent in the atmosphere that their effects are difficult to discern in isolation from natural clouds. But an unusual situation arose in the wake of the terrorist attacks of 11 September 2001, when all commercial flights in the United States were grounded for several days. The skies were suddenly contrail-free for the first time since the 1920s, allowing a controlled study to be made of a temporarily unaviated atmosphere. The results, according to a comparison of nationwide temperature records, were slightly warmer days and slightly cooler nights than was usual for that time of year, the normal night/day temperature range having increased by 1.1°c. According to the climate scientists who worked on the data, this was a likely consequence of additional sunlight reaching the surface

Artificial rainclouds
created for the
opening ceremony
of the London 2012
Olympic Games in
case the real thing
failed to appear.

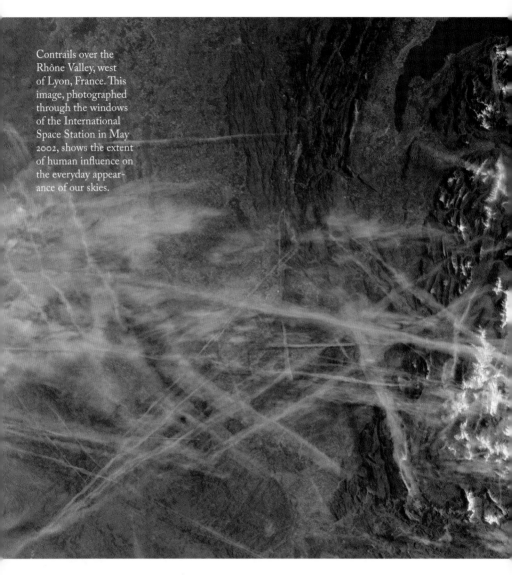

Contrails over the Rhône Valley, west of Lyon, France. This image, photographed through the windows of the International Space Station in May 2002, shows the extent of human influence on the everyday appearance of our skies.

by day, and additional radiation escaping at night through the contrail-free skies.[14] This might appear counter-intuitive, for surely the kind of cirriform clouds created by the spreading of aircraft contrails are straightforward warming clouds, the kind that admit sunlight in from above, while redirecting radiation down to the surface. An absence of contrails ought, therefore, to

have had an overall *cooling* effect. But contrails are a lot more complicated than that, because when they are in their initial, water droplet stage they are far denser than natural cirrus clouds, since they are created from two distinct sources of vapour: the moisture emitted by the aircraft's exhaust plus the moisture already in the atmosphere. At first, this opaque contrail behaves more like a white, low-level cloud, reflecting sunlight back into space, and exerting a short-term localized cooling effect. But as persistent contrails begin to spread, they thin out into recognizably cirriform cloud layers, their supercooled water droplets having frozen into the kind of tiny ice crystals associated with cirrostratus. Their overall effect thus reverts to a warming one, consistent with the observed behaviour of natural cirriform clouds.

The picture is complicated further, however, by the time of day or night that the contrails form and spread. If contrails spread during the early morning or late evening, they can exercise a slight cooling effect, due to the angle at which sunlight tends to reflect off the ice crystals rather than penetrate through to the ground. At night, by contrast, all clouds, including contrails, can only exert a warming effect, since there is no incoming sunlight to reflect back into space: clear nights are always colder than cloudy ones. Any increase in night flights (an increase that is already under way) is therefore likely to lead to slightly raised temperatures on the ground. In fact, the warming effects of the predicted increase in contrail production, particularly those associated with the rise in night flights, have been projected to be in the region of a 0.2°c to 0.3°c hike per decade in the United States alone (a figure that does not include other warming effects associated with increased aviation, such as CO_2 emissions and local ozone formation).[15]

There is a lot still to learn about the complex behaviour of these high-altitude anthropogenic clouds. Whether planes of the future will need to change the routes or altitudes at which they fly in order to lessen or modify contrail formation is a matter of ongoing speculation. Research has shown that, on average, only 7 per cent of the total distance flown by an aircraft is through the kind of air in which long-lasting contrails form, so adjusting routes to avoid those easily identifiable areas could reduce the creation of warming clouds. A flight from London to New York, for example, could be flown at a slightly higher altitude over the Atlantic, adding only around 20 km to the length of the flight, while significantly reducing the formation of spreading contrails, whose warming effects far exceed those of the same aircraft's CO_2 emissions.[16]

A view from under the flight path: contrails over east London, July 2015.

Fallstreak holes (also known as hole-punch clouds), photographed from the Cold Springs Bridge, California, March 2015.

Contrails are not the only visible signs of commercial aviation. When aircraft pass through layers of cloud they can leave linear gaps in their wake, known as distrails (short for 'dissipation trails'), as well as a series of telltale holes and streaks known variously as hole-punch clouds, fallstreak holes or canal clouds. Although the physics of fallstreaks are not yet completely understood, they are known to be created by the sudden freezing of an isolated patch of supercooled cloud, which then falls away in a shower of ice crystals, leaving a visible gap behind. ('Supercooled' clouds are composed of water droplets that remain in a liquid state at temperatures well below freezing.) Aircraft exhaust is likely to be the main cause of hole-punch clouds, since supercooling occurs when there are not enough freezing nuclei available for airborne water droplets to turn into ice, and aircraft exhaust is an abundant source of such nuclei. Research published in the journal *Science* in 2011 described the creation of fallstreak holes as a form of 'inadvertent cloud seeding', produced by the spontaneous freezing of cloud droplets as air flows around aircraft propeller tips or over jet aircraft wings.[17] Such holes tend to form when a plane is climbing or descending through clouds at a shallow angle, leaving a kind of spatial shadow in its wake. They are unlikely to induce rain, however, as the falling ice rarely gets far, usually melting and then evaporating long before it reaches the ground.

Ship tracks – the maritime equivalent of aircraft contrails – are a less familiar side effect of mass transportation, but their atmospheric impact is almost as great. Like contrails, ship tracks are linear clouds seeded by microscopic particles (mainly sulphates) present in a ship's exhaust. Because the exhaust particles are more abundant than natural airborne particles (usually sea salt), they generate smaller and more abundant cloud droplets; as a consequence, ship tracks tend to be whiter and more reflective than natural maritime clouds, thereby reducing the amount of incoming sunlight over the ocean surface. There is an irony in the fact that the dirtiest, most polluted clouds have the highest reflective indexes, and so are better at offsetting the greenhouse effect by redirecting sunlight back into space. Even though sea surface temperatures are continuing to rise, dirty clouds, such as ships' tracks, exert a measurable cooling effect and, predictably, there have been calls to generate artificially brightened maritime clouds over large areas of ocean as a way of reducing increased surface warming. But, as is always the case with large-scale geo-engineering projects, the law of unintended consequences needs to be kept in mind, and so far it isn't clear what other environmental effects such artificial cloud-mirrors could have. In the meantime, the World Meteorological Organization is preparing to add the term *homogenitus* ('man-made') to the official cloud classification as a way of denoting the increasing array of anthropogenic clouds that are being written across our seas and skies.

Among the most common of these are industrial cumulus (also known as 'fumulus') clouds, which form above industrial cooling towers. The warm, moist air from the towers rises and rapidly cools, condensing, like steam above a boiling kettle, into low-lying cumuliform clouds, the bases of which can sometimes

An industrial cumulus ('fumulus') cloud forming above a geothermal plant in southwest Iceland.

be only a few hundred metres above the ground. If moisture is already present in the atmosphere, fumulus clouds can grow to considerable sizes and drift for many kilometres across the surrounding area, sometimes even going on to produce rain. Fire-heated clouds can also produce rain – and sometimes even flashes of lightning – as is often seen above forest fires, when the intense heating of moisture-laden air produces dark, smoky cumulus clouds, known as pyrocumulus. (Some have been known to grow so vast that they are referred to, unofficially, as pyro-cumulonimbus.) They often form naturally, above wildfires and volcanic eruptions, but can also be artificially produced by large-scale agricultural or industrial combustion.

Overleaf: Ship tracks over the northern Pacific Ocean: the maritime equivalent of aircraft contrails.

Regular visitors to airshows will be familiar with the variety of short-lived clouds and cloud effects that can be produced by aerial manoeuvres. Sudden ascents and descents can pressurize the

Large pyrocumulus cloud cooked up by the deadly Station Fire on the slopes of Mount Wilson, California, in August 2009.

air around an aircraft's wings, cooling and condensing ambient water vapour into visible clouds that last only a second or two, but can be an impressive sight, shrouding or trailing the speeding plane for a magical moment, as though conjured out of thin air.

Cloud architecture

When the Victorian meteorologist William Clement Ley invoked 'the dim outlines of a future science of cloud-land', he could hardly have imagined the kind of atmospheric engineering that has become a recurring motif in recent artistic and architectural practice.[18] The Japanese artist Fujiko Nakaya made her name in the 1970s for her innovative 'fog sculptures', created using hundreds of mist-generating water nozzles powered by high-pressure motorized pumps. Each of her installations has been site-specific, its bespoke design based on meteorological records of local wind and weather conditions. One of her most celebrated pieces, *Cloud Parking* (2011), allowed visitors to walk through a bank of artificial stratus cloud created on the roof of a multi-storey car park in Linz, Austria, while an earlier piece, *Fog Sculpture #08025* (1998) is now a permanent installation at the Guggenheim Museum Bilbao, where it surprises visitors by intermittently appearing and dissipating over the gallery's

Steam emerging from an Icelandic geyser creates ground-level cloud that rises and spreads to form atmospheric cloud.

A short-lived sonic boom cloud, created by a Hornet fighter jet breaking the sound barrier over the Pacific Ocean, 7 July 1999.

Fujiko Nakaya's *Fog Sculpture #08025*, 1998, an artificial cloud created with high-pressure water nozzles, Guggenheim Museum, Bilbao. Nakaya's father was the physicist Ukichiro Nakaya, famous for creating the world's first artificial snow flakes.

Spearing through a
layer of stratocumulus
cloud, the manned
Space Shuttle
Endeavour blasts
off on its final
mission in May
2011, leaving a
dramatic vertical
contrail in its wake.

Wingtip vortices trailing from an F-15E Strike Eagle as it disengages from mid-air refuelling, 2003. Vortices are rotating streams of air left behind a wing as it generates lift, and are sometimes made visible by the sudden condensation of ambient moisture in the atmosphere.

ornamental pond like an atmospheric ha-ha. In March 2017 a new version of the sculpture formed the inaugural installation of the Switch House extension, Tate Modern, London.

Nakaya acted as a consultant for Elizabeth Diller and Ricardo Scofidio's *Blur* building, the media pavilion for the 2002 Swiss National Expo at Yverdon-les-Bains, which took the form of an inhabitable cloud that appeared to hover over Lake Neuchâtel, like the floating Cloud City from *The Empire Strikes Back* (dir. Irvin Kershner, 1980). The dimensions of the structure varied according to wind conditions: on a calm day it measured around 90 metres (295 ft) across and some 60 metres (197 ft) deep, but even in the stillest conditions it was always in flux, growing and dissipating in response to fluctuations of internal and external weather. The building's curved frame was fitted with over 30,000 fog nozzles that pumped lake water into a fine mist, the contours of which were controlled by a computerized weather system, like the cloud factory imagined in Timothy Donnelly's poem sequence *The Cloud Corporation* (2010):

Fans conveying clouds through aluminum ducts
can be heard from up to a mile away, depending on
air temperature, humidity, the absence or presence

of any competing sound, its origin and its character.
It is no more impossible to grasp the baboon's
full significance in Egyptian religious symbolism

than it is to determine why clouds we manufacture
provoke in an audience more positive, lasting
response than do comparable clouds occurring in nature.[19]

Visitors to the *Blur* building were issued with white interactive rainwear, dubbed 'braincoats', before ascending the stairs to the 'Angel Bar' in the clouds, where a selection of waters from around the world were waiting to be sampled. For many of the journalists stationed in this futuristic 'wonder-cloud', the experience was both disorientating and sublime. As Lawrence Weschler

commented, 'You're never going to top the primordial sensation of walking atop water – and into a cloud!' – although Antony Gormley's *Blind Light* (2007) arguably went further, enclosing a brightly lit cloud inside a walk-through glass vitrine.[20] Inside the 10 m x 10 m box was a cooled, watery mist, which, under the impact of 7,000 lux of intense fluorescent light, gave a visibility of less than an arm's length. One could, and did, get temporarily lost in its 90 per cent humidity. The intended effect of Gormley's 'bright, cuboid cloud' was to overwhelm the senses, as though one had walked into a cloud, literally and figuratively, entering a cold, damp, unsettling world of enveloping isolation.[21] As designer Ben Rubin observed, in relation to the *Blur* building, 'the luminosity of the cloud presents a signal-to-noise problem: the cloud is pure visual noise, undifferentiated white light, lots of it, scattering in every direction', a disorientating quality that Gormley successfully recreated in his glowing walk-in cloud chamber.[22]

An F/A-18F Super Hornet pulls a high-G manoeuvre, causing large amounts of condensed water to form a short-lived cloud above its wings, September 2004.

One of Gormley's earlier creations, *Quantum Cloud* (1999), raised similar questions about the status of a cloud as a tangible yet indeterminate object. The 30-m-high steel sculpture, which is now a permanent fixture on the Greenwich Peninsula opposite London's Millennium Dome, was designed using fractal software, which plotted a nebulous ('random walk') arrangement of the tetrahedral steel units from which the piece is constructed. Like a natural cloud, the sculpture has diffuse edges, lacking a clearly definable boundary, although a residual outline of a human body can be detected at the centre. 'A dematerialised monument', as Gormley has described it, the computer-generated *Quantum Cloud* is truly a cloud of our time.[23]

The opening lines of Wordsworth's sonnet, 'Upon the Sight of a Beautiful Picture, Painted by Sir G. H. Beaumont, Bart' (1811), offer a meditation on art's capacity to give permanence to

Antony Gormley, *Quantum Cloud*, 1999, installed on the bank of the Thames opposite London's Millennium Dome.

fleeting events: 'Praised be the Art whose subtle power could stay/ Yon cloud, and fix it in that glorious shape.'[24] The sentiment applies as much to photography as to easel painting, and especially to the kind of 'blink-and-you'll-miss-it' artworks with which the Dutch cloud sculptor Berndnaut Smilde has made his name. The first of Smilde's ongoing *Nimbus* series, featuring eerily backlit indoor clouds, was installed in a former chapel in 2012 — if 'installed' is the right word for an airborne apparition lasting less than ten seconds. Since then the artist has created and photographed his ephemeral clouds in a variety of spaces around the world, from coal mines to castles and cathedrals. Smilde's creations require detailed preparation to create the environmental conditions in which they can live long enough to be captured on camera. Each interior space needs to be cold and damp, with limited air movement, otherwise the fog from the smoke machine would fail to coalesce into a cloud. Smilde mists the air to a high degree of humidity, using handheld plant spray bottles,

Bringing clouds indoors: the view inside Willett's Amazing Camera Obscura, September 2015.

before releasing a burst of 'fog' (actually a 'vaporized glycol-based compound') from a smoke machine off-camera. The ambient moisture sticks to the 'fog', preventing it from drifting apart, allowing a white cloud to form and float indoors for a few eerie seconds, before it vanishes into the air. 'I really like the impermanent aspect of it', observes Smilde. 'It's about this very short moment in time, a specific location; it's almost like a memory of this cloud that happened there.'[25]

A more permanent way of incorporating clouds into the built environment is to create the kind of oculus or viewing portal that the American light artist James Turrell has installed in more than eighty galleries and other spaces around the world as part of his *Skyspace* series. *Skyspace I* (1975) was an opening cut into the roof of the Villa Panza in Varese, Italy, which acted as a portal through which clouds and weather could be viewed from the dimly lit sanctuary of the gallery space. Turrell (who, like Luke Howard, is a practising Quaker) has installed Skyspaces in a number of Quaker meeting houses, including a piece entitled *Greet the Light* (2013) at the Chestnut Hill Friends Meeting in Philadelphia, in which the framing of the sky reflects the contemplative nature of the prayer meetings. As the theologian Jeffrey Kosky has observed, 'You do not look at what [Turrell] has made, for what he has made is just the edge circling around what is not made: the sky that his work is always *about*.'[26] Turrell's best-known work, the ongoing Roden Crater Project (1975–), is, in effect, a giant Skyspace cut into an extinct volcanic cone in the Painted Desert some 80 km (50 miles) northeast of Flagstaff, Arizona. At night, the terraformed Crater transforms into a spectacular astronomical observatory, but in daylight hours it is the clouds that are the stars of the show, with frequent displays of stacked lenticular clouds (*'piles d'assiettes'*) that form above the San Francisco Peaks. An aperture in one of the crater's numerous excavated chambers is fitted with a camera obscura lens that, during the day, projects cloud patterns onto a circular floor: 'first the low cumulus clouds, then the higher cirrus clouds, and sometimes the nacreous clouds very high up in the stratosphere'.[27] Roden Crater is

an eye, something that is itself perceiving ... it is changed
by the action of the sun, the moon, the cloud cover, or what
day or what season you're there, and it keeps changing.
When you're there, it has visions, qualities, and a universe
of possibilities.[28]

Clouds and climate change

Clouds and cloud behaviour play a valuable role in indicating
short-range weather conditions, but when it comes to predicting
longer-term climatic changes, they are entirely unknown quan-
tities. For despite the scientific consensus on the reality of global
climate change, the subject remains riddled with uncertainties,
among the most pressing of which concerns the likely role that
clouds will play in shaping future conditions on earth. Will
clouds turn out to be agents of global warming, veiling us in an
ever-thickening blanket of greenhouse gas emissions (of which
water vapour is among the most potent), or will they end up
saving the day by reflecting ever more sunlight back into space?
These, it turns out, are far from simple questions, and as the
Fourth Assessment Report (2007) of the Intergovernmental
Panel on Climate Change made clear, clouds and cloud behaviour are the true wild cards in the game of forecasting future
climate:

> A change in almost any aspect of clouds, such as their type,
> location, water content, cloud altitude, particle size and
> shape, or lifetimes, affects the degree to which clouds warm
> or cool the Earth. Some changes amplify warming while
> others diminish it. Much research is in progress to better
> understand how clouds change in response to climate
> warming, and how these changes affect climate through
> various feedback mechanisms.[29]

The Fifth Assessment Report (2013) supplied a wealth of
technical detail gleaned from this ongoing research, but still
concluded that 'clouds and aerosols continue to contribute the

largest uncertainty to estimates and interpretations of the earth's changing energy budget'.[30] Clouds, as always, confound our understanding, and much of the research and modelling yields apparently contradictory results. So while some climate models suggest that continued surface warming will see an increase in water vapour rising from the oceans, leading to an overall increase in cloud formation, others suggest that, in warmer latitudes, an increase in the water vapour content of our atmosphere would see large convective cumuliform clouds building up and raining themselves out far quicker than they do at present, leading to a net *decrease* in the earth's total cloud cover. Low-level stratiform clouds, meanwhile, tend to shield the earth from incoming solar radiation, but modelling has indicated that such clouds are more likely to dissipate in warmer conditions, thus allowing the oceans to heat up even more and causing yet further stratiform cloud loss.[31] Scientists currently have no idea which outcome is the more likely, nor do they really know the kind of long-term influences that either would be likely to have. Even if it is assumed that overall cloud cover will increase as the surface of our planet continues to warm, it remains unclear what kind of clouds (and thus what kind of feedback scenarios) are likely to predominate. For instance, high, thin cirriform clouds, such as cirrostratus, tend to have an overall warming effect, as they admit a lot of shortwave radiation in from above (in the form of sunlight during the day), while intercepting longwave back-radiation (warmth reflected from the sunlit ground) and despatching it back down to earth. Any increase in cirrostratus cloud cover (including from spread-out aircraft contrails) would therefore result in the addition of yet another warming mechanism to our climate. In contrast, however, bright, dense clouds, such as cumulus congestus, serve to cool the earth by reflecting incoming sunlight back into space by day. At night, these same clouds can exert a slight warming effect, by absorbing or reflecting back radiation, but their overall influence is a cooling one, especially when their summits grow reflectively dense and white. Currently, around 20 per cent of the solar energy that reaches the earth is reflected back into space from clouds – the largest heat transfer on the planet.

What looks like a layer of natural cirrostratus cloud has in fact been formed by aircraft contrails that have spread out and merged some 10 km (30,000 ft) above London. Such man-made skies will become an increasingly familiar sight as the world's flight paths grow busier.

So, in theory, an increase in high, thin layers of cloud would amplify the global warming effect, while an increase in low, dense, puffy clouds would exert a contrary cooling influence. In reality, of course, things are never so simple or predictable and, as has been seen throughout this book, clouds have a habit of behaving in complex and surprising ways. Take lightning, for example. The earth experiences more than a billion lightning flashes per year, but it has been known for some time that our warming climate is producing ever more lightning strikes, with growing numbers of fatalities recorded in developing countries. Research published in the journal *Science* in November 2014 showed that incidents of lightning increase proportionately to the temperature of the atmosphere, due to the increased heat energy available to fuel large convective storm clouds. Since lightning activity increases by around 12 per cent per 1°c of atmospheric warming, lightning strikes are currently on track

Night-time thunderstorm illuminated by in-cloud lightning, Nebraska, June 2010. The number of lightning strikes is predicted to rise by around 50 per cent by the end of this century.

to increase by around 50 per cent by the end of the twenty-first century.[32]

Rain, too, is full of surprises, being a sort of refrigerated soup concocted from water, dust, spores and aerosols and all manner of things that you find in clouds, including viruses and bacteria. In the 1930s the pioneer aviator Charles Lindbergh devised a tubular contraption called a 'sky hook' that he used to collect fungi and pollen from the teeming atmosphere. Some of what he found in the clouds came as a revelation. Many species of bacteria, for example, have made their way up into clouds, where they act as freezing nuclei, precipitating ice formation at higher than normal temperatures. Leaf fragments have also been found suspended in clouds and acting as freezing nuclei, although the ice crystals form not on the leaf fragments themselves, which are too large, but on bacteria lifted up with the leaves and on which they feed and reproduce, creating little self-sufficient bacterial colonies high up in the clouds.

The idea of bacterial clouds might sound like something from science fiction, but the most pressing question they raise concerns changes to the rainmaking nature of these so-called 'living clouds', which meteorologists suspect of producing rain more readily than clouds without bacteria. As William Bryant Logan observes,

> Bacteria do double work in the terraqueous machine. While underground, they compost the dead, protect roots, and fix from the air the nitrogen without which not a single amino or nucleic acid can be made. Aloft in the clouds, they create condensation nuclei, and thus affect the amount and frequency of rain and snow.[33]

Strange as it may seem, future weather forecasts may well depend on knowing the bacterial content of clouds as a means of predicting rain with greater precision.

A more distant question for the future concerns clouds as potential sources of energy. When vapour condenses into water, it releases a great deal of thermal energy, and that energy warms

The 'Baker' explosion, Bikini Atoll, 25 July 1946. The exterior cloud was a more diffused condensation cloud than the classic 'mushroom cloud' associated with some of the later bomb tests.

the air around it, creating bursts of buoyant turbulence inside clouds, which is why most convective clouds have roiling, uneven tops. Even a small cumulus cloud generates a surprising amount of heat energy – enough to power a family home for several years – while a million-tonne cumulonimbus cloud contains more thermal energy than a hundred nuclear warheads. The challenge, of course, would be to harness that energy: one idea involves the use of cloud-to-ground electrodes which would draw down atmospheric charge in the form of an electric current, but such technology, even if it were remotely plausible, is still a long way off.

Sunlight streams through the summits of cumulus castellanus clouds in Félix Bracquemond's atmospheric *The Storm Clouds*, early 1880s, etching.

'Clouds always tell a true story,' as Ralph Abercromby observed in 1887, 'but one that is difficult to read.'[34] And though Abercromby was referring to the problem of understanding the relationship between clouds and weather, his comment can be applied just as aptly to the relationship between clouds and climate. As has been seen, there are many interconnected factors which make the life stories of clouds such a challenge to read and understand, among the most daunting of which is the fact that, as our climate warms, the atmosphere correspondingly reorganizes itself in ways that could either amplify or mitigate the original warming. Our changing climate has the capacity to alter the day-to-day behaviour of clouds and weather in all kinds of unpredictable ways. Such is the uncertainty over the likely influence of global feedback mechanisms, especially those involving clouds, that the only thing that can be said with confidence is that clouds could increase future warming, decrease future warming, or exert an effect somewhere in between – including having no effect at all. In short, we have no way of knowing what is really going to happen to our increasingly anthropogenic atmosphere, and just as, in centuries past, clouds were employed as ready metaphors of doubt and uncertainty, it looks as if they will continue to be so for centuries to come.

APPENDIX: CLOUD SPECIES AND VARIETIES

The current cloud classification, as administered by the World Meteorological Organization in Geneva, is based on a tripartite altitude structure. The ten main cloud genera, with their international abbreviations and the dates of their agreed definition or redefinition, are as follows:

HIGH CLOUDS, bases usually above 6 km (> 20,000 ft):
Cirrus, Ci (Howard, 1803)
Cirrocumulus, Cc (Howard, 1803; Renou, 1855)
Cirrostratus, Cs (Howard, 1803; Renou, 1855)

MEDIUM CLOUDS, bases usually between 2 and 6 km
(6,500 and 20,000 ft):
Altocumulus, Ac (Renou, 1870)
Altostratus, As (Renou, 1877)
Nimbostratus, Ns (International Commission
for the Study of Clouds, 1930)

LOW CLOUDS, bases usually below 2 km (< 6,500 ft):
Stratocumulus, Sc (Kaemtz, 1840)
Stratus, St (Howard, 1803; Hildebrandsson and Abercromby, 1887)
Cumulus, Cu (Howard, 1803)
Cumulonimbus, Cb (Weilbach, 1880)

Most of the cloud genera (with the exception of altostratus and nimbo-stratus) are divided into species, with Latin names based on peculiarities in their shapes or differences in their internal structures. The full list of cloud species, with their abbreviations and meanings, is as follows:

Calvus, cal
('bald'), applied to cumulonimbus clouds without an icy anvil

Capillatus, cap
('hairy'), applied to cumulonimbus clouds with a
fibrous anvil or plume

Castellanus, cas
('castle-like'), applied to stratocumulus, altocumulus,
cirrus or cirrocumulus clouds with turret-like or crenellated
protuberances in their upper regions

Congestus, con
('piled up'), applied to the kind of large and rapidly
growing cumulus cloud whose summit resembles a cauliflower

Fibratus, fib
('fibrous'), applies mainly to thin cirrus or cirrocumulus
clouds with straight or irregularly curved filaments

Floccus, flo
('tufted'), applied to small altocumulus, cirrus or cirro
cumulus clouds

Fractus, fra
('broken'), ragged shred of stratus and cumulus cloud

Humilis, hum
('low'), applied to cumulus clouds of slight vertical extent

Lenticularis, len
('lens-like'), applied mainly to stratocumulus or
altocumulus clouds in the shape of an elongated lens or almond

Mediocris, med
('medium'), describes cumulus clouds of moderate
vertical extent

Nebulosus, neb
('misty'), describes thin, indistinct layers of stratus or
cirrostratus clouds

Spissatus, spi
('thick'), applies to unusually dense cirrus clouds

Stratiformis, str
('layered'), applies to stratocumulus, altocumulus and,
occasionally, cirrocumulus clouds that have spread into a horizontal layer

Uncinus, unc

('hooked'), applied to cirrus clouds shaped like a comma
or hook

Volutus, vol

('rolled'), currently under consideration for inclusion in
the next edition of the *International Cloud Atlas* (2017)

Certain features of clouds' appearance or degree of transparency have led
to the subdivision of a number of species into varieties. The full list of cloud
varieties, with their abbreviations and meanings, is as follows:

Asperitas, asp

('roughness'), applied to wavelike structures in the
underside of an extensive undulatus cloud

Cumulogenitus, cugen

('made from cumulus'), applies to a stratocumulus
or altocumulus cloud that is formed by the spreading out of
cumulus

Duplicatus, du

('doubled'), applies to layers of stratocumulus,
altocumulus, altostratus, cirrus or cirrocumulus clouds that are
formed at slightly different levels, sometimes partly merged

Homogenitus, hom

('man-made'), applies to anthropogenic clouds such
as contrails or industrial cumulus ('fumulus') clouds

Intortus, in

('twisted'), describes cirrus clouds with irregularly curved
or tangled filaments

Lacunosus, la

('with gaps'), applies to reticulated patches of altocumulus
and cirrocumulus clouds

Opacus, op

('opaque'), applied to stratus, stratocumulus, altostratus or
altocumulus clouds in an extensive patch or layer, the greater part
of which is sufficiently opaque to mask the sun or moon

Perlucidus, pe

('allowing light through'), applies to extensive layers of
stratocumulus or altocumulus clouds, with thinner patches that
allow the sun or moon to be seen; the variety perlucidus is often
observed in combination with the translucidus or opacus varieties

Radiatus, ra
('radiated'), applied to broad parallel bands of stratocumulus, cumulus, altocumulus, altostratus or cirrus clouds that
appear to converge towards a point on the horizon

Translucidus, tr
('translucent'), applies to stratus, stratocumulus, altostratus or altocumulus clouds in an extensive patch or layer, the
greater part of which is sufficiently translucent to reveal the
position of the sun or moon

Undulatus, un
('wavy'), applies to patches or layers of cloud with
noticeable waves or undulations

Vertebratus, ve
('like vertebrae'), usually describes a cirrus cloud whose
filaments form a fishbone pattern

TIMELINE

c. 420 BC	Aristophanes, *The Clouds*
c. 340 BC	Aristotle, *Meteorologica*
c. AD 1360–90	Anon., *The Cloud of Unknowing*
1563	William Fulke, *A Goodly Gallerye*
1637	René Descartes, *On Meteors*
1665	Robert Hooke attempts to establish an agreed vocabulary for clouds and weather
1703	'The Great Storm', 26–7 November
1780s	The German Societas Meteorologica Palatina proposes a descriptive cloud classification
1802	Jean-Baptiste Lamarck proposes a French-language cloud classification, arranged by altitude
1802	December: Luke Howard delivers his landmark lecture on the modifications of clouds
1803	Howard publishes his cloud nomenclature in the *Philosophical Magazine*
1815	Howard's *Essay on the Modifications of Clouds* published in French and German translations
1817	J. W. Goethe, *Howards Ehrengedächtnis*
1820	Shelley, 'The Cloud' and 'Ode to the West Wind'
1821–2	Landscape painter John Constable produces more than a hundred *plein-air* cloud studies on Hampstead Heath
1823	Founding of the London (later the Royal) Meteorological Society
1840	Ludwig Kaemtz reclassifies *Cumulo-stratus* as *Stratocumulus*
1855	Émilien Renou proposes two new cloud genera, *Altocumulus* and *Altostratus*, officially added to the cloud classification in 1870 and 1877 respectively

1862	Balloonists James Glaisher and Henry Coxwell become the first people to enter the stratosphere
1863	Andrés Poey adds the variety *Cumulus fractus* to the cloud classification
1865	Expanded 3rd edition of Luke Howard's *Essay on the Modifications of Clouds*, with newly commissioned illustrations
1880	Amateur meteorologist Philip Weilbach classifies *Cumulonimbus*
1884	April: the first documented photograph of a tornado taken in Kansas
1887	Hugo Hildebrandsson and Ralph Abercromby propose an international cloud classification, based on Howard's nomenclature
1889	Claude Debussy's nocturne *Nuages* ('Clouds') premiered in Paris
1890	Hildebrandsson and two others publish the first multilingual, illustrated *Cloud Atlas*
1896	The International Year of Clouds: the first *International Cloud Atlas* published
1905	*Lenticularis* added to the cloud classification
1922	Photographer Alfred Stieglitz makes the first of more than two hundred cloud studies, later known as *Equivalents*
1947	Launch of Project Cirrus, a u.s.-led cloud and weather modification programme
1951	*Intortus* added to the wmo's cloud classification
1952	Psychoanalyst Wilhelm Reich invents the 'cloud-buster'
1956	Fifth edition of the *International Cloud Atlas*, the first to be be divided into two volumes – text and plates – for ease of translation
1970	Artist Fujiko Nakaya makes her first fog sculpture, Osaka, Japan
1974	Artist James Turrell installs his first *Skyspace*, Varese, Italy
1995	Seventh edition of the *International Cloud Atlas*
1999	Antony Gormley's *Quantum Cloud* permanently installed next to the River Thames, overlooking the Millennium Dome
2002	Diller & Scofidio's *Blur* building, Yverdon-les-Bains, Switzerland
2004	Cloud Appreciation Society founded by Gavin Pretor-Pinney
2008	Cloud seeding deployed during the Beijing Olympics
2013	The Fifth Assessment Report of the ipcc states that clouds 'continue to contribute the largest uncertainty' to projections of future climate

2015   *Asperitas; Volutus; Homogenitus* added to the WMO's cloud classification, the first new cloud terms since 1951

2017   Publication of new, online edition of the *International Cloud Atlas*

REFERENCES

Introduction: 'Airy Nothings'

1 Aristophanes, *Lysistrata and Other Plays*, trans. Alan H. Sommerstein (Harmondsworth, 1973), p. 125.
2 Ibid., p. 126.
3 Mary Jacobus, 'Cloud Studies: The Visible Invisible', *Gramma: Journal of Theory and Criticism*, 14 (2006), p. 221.
4 Cited in Jack Rochford Vrooman, *René Descartes: A Biography* (New York, 1970), pp. 123–4.
5 Tao Hongjing, 'Reply to the Imperial Inquiry', cited in Pao K. Wang, *Physics and Dynamics of Clouds and Precipitation* (Cambridge, 2013), p. ix.
6 Ernst Bloch, 'Better Castles in the Sky', in *The Utopian Function of Art and Literature: Selected Essays*, trans. Jack Zipes (Cambridge, MA, 1988), p. 175.
7 '20', in Wendy Mark and Mark Strand, *89 Clouds* (New York, 1999).
8 Alexandra Harris, *Weatherland: Writers and Artists Under English Skies* (London, 2015), pp. 11–12. In her book, Harris also cites a riddle (answer: 'cloud'), from the *Enigmata* of the Anglo-Saxon monk Aldhelm:

> In flight I vary my hue, leaving heaven and earth behind: there is no place on earth for me, none in the region of the skies. No other creature fears an exile of such cruelty; but I make the world grown green with my moist drops.

1 Clouds in Myth and Metaphor

1 Kevin Crossley-Holland, trans., 'The Lay of Alvis', in *The Penguin Book of Norse Myths: Gods of the Vikings* (London, 1993), p. 144.

2 Cited in Gustav Hellmann, 'The Dawn of Meteorology', *Quarterly Journal of the Royal Meteorological Society*, XXXIV (1908), pp. 223–4.

3 In E.T.C. Werner, *Myths and Legends of China* (New York and London, 1922), p. 303.

4 Ralph Abercromby, 'Cloud-land in Folk-lore and Science', *Folk-lore Journal*, VI (1888), p. 102.

5 Frank G. French, ed., *The Sutra of Bodhisattva Ksitigarbha's Fundamental Vows*, 2nd edn (Taipei, 2004), p. 11.

6 Crossley-Holland, *The Penguin Book of Norse Myths*, p. 5. In Thomas Mann's novella *Death in Venice* (1912), a fateful day in the cholera-struck city begins with 'troops of small feathery white clouds ranged over the sky, like grazing herds of the gods'.

7 Ted Hughes, *Tales from Ovid: Twenty-four Passages from the Metamorphoses* (London, 1997), pp. 98–9.

8 Cited in Richard Inwards, *Weather Lore: A Collection of Proverbs, Sayings, and Rules Covering the Weather*, 4th edn (London, 1950), p. 137.

9 *Religion and American Cultures: Tradition, Diversity, and Popular Expression*, ed. Gary Laderman and Luis León, 2nd edn (Santa Barbara, CA, 2014), vol. I, p. 236.

10 Ibid., p. 237.

11 See Walter Scott, *Letters on Demonology and Witchcraft*, 2nd edn (Edinburgh, 1831), p. 391.

12 See Thomas Glick et al., eds, *Medieval Science, Technology, and Medicine: An Encyclopedia* (New York, 2005), pp. 342–5. Theophrastus translation from Inwards, *Weather Lore*, p. 125.

13 Cited in Inwards, *Weather Lore*, p. 123.

14 *A Goodly Gallerye: William Fulke's Book of Meteors (1563)*, ed. Theodore Hornberger (Philadelphia, PA, 1979), pp. 92–3.

15 John Claridge, *The Shepherd of Banbury's Rules to Judge of the Changes of the Weather, Grounded on Forty Years Experience* (London, 1744), p. ii.

16 Ibid., p. 5.

17 *The Arabian Nights: Tales of 1001 Nights*, trans. Malcolm C. Lyons, 3 vols (London, 2008), II, pp. 343–4. The story covers nights 473 and 474.

18 W. G. Sebald, *The Rings of Saturn*, trans. Michael Hulse (London, 1998), p. 241.

19 *A Goodly Gallerye*, p. 79.

20 In *Poems and Prose of Gerard Manley Hopkins*, ed. W. H. Gardner (Harmondsworth, 1953), p. 109.

21 G. M. Hopkins, 'Shadow-beams in the East at Sunset', *Nature*, XXIX (1883), p. 55.

22 'The Apparition of the Brocken', in *The Collected Writings of Thomas De Quincey*, ed. David Masson (Edinburgh, 1889), vol. I, pp. 51–4.

23 'A Description of the Province of Quito, and Other Parts of Peru',
    Gentleman's Magazine, 19 (1749), pp. 216–17.
24 James Hogg, *The Private Memoirs and Confessions of a Justified
    Sinner*, ed. John Carey (Oxford, 1969), pp. 39–41.
25 James Hogg, 'Nature's Magic Lantern', in *Tales and Sketches by the
    Ettrick Shepherd* (London, 1878), p. 353.
26 Meiko O'Halloran, *James Hogg and British Romanticism:
    A Kaleidoscopic Art* (London, 2015), pp. 199–200.
27 In Walt Whitman, *The Complete Poems*, ed. Francis Murphy
    (London, 2004), pp. 286–7.
28 James Glaisher et al., *Travels in the Air* (London, 1871), p. 29.
29 Cited in Helen Sawyer Hogg, 'Le Gentil and the Transits of Venus,
    1761 and 1769', *Journal of the Royal Astronomical Society of Canada*,
    XLV (1951), p. 132.
30 *The Diary of John Evelyn*, ed. E. S. de Beer, 6 vols (Oxford, 1955), II,
    pp. 207–8.
31 William Wordsworth, *The Prelude: A Parallel Text*, ed. J. C. Maxwell
    (Harmondsworth, 1971), p. 513 (XIV, 41–7).
32 Thomas Baldwin, *Airopaidia; Containing the Narrative of a Balloon
    Excursion from Chester, &c.* (Chester, 1786), pp. 53–4.
33 Glaisher et al., *Travels in the Air*, p. 49.
34 Richard Holmes, *Falling Upwards: How We Took to the Air*
    (London, 2013), p. 247.
35 Stéphane Audeguy, *The Theory of Clouds*, trans. Timothy Bent
    (New York, 2007), p. 63.
36 Saul Bellow, *Henderson the Rain King* (New York, 1959), p. 280.
37 Malka Marom, ed., *Joni Mitchell: In Her Own Words* (Toronto,
    2014), p. 35. Mitchell evidently has a strong visual memory, for
    although the line appears on p. 280 of the novel, she remembered it,
    decades later, as p. 28.
38 Derek Walcott, *White Egrets* (London, 2010), p. 89.

2 The Natural History of Clouds

1 Charles M. Schulz, *You Can Do It, Charlie Brown* (New York, 1963),
   p. 14. A cartoon in the same collection features a ground-hugging
   cloud 'that's afraid of heights'.
2 William Clement Ley, 'Clouds and Weather Signs', in *Modern
   Meteorology: A Series of Six Lectures* (London, 1879), p. 106.
3 James Glaisher et al., *Travels in the Air* (London, 1871), p. 22.
4 H. Howard Frisinger, 'Aristotle and his "Meteorologica"', *Bulletin of
   the American Meteorological Society*, LIII (1972), p. 636.
5 Ovid, *Metamorphoses*, trans. Mary M. Innes (Harmondsworth, 1955),
   p. 30.

6 Seneca, *Naturales quaestiones*, trans. T. H. Corcoran (London, 1971), vol. II, p. 273.

7 Translation cited in Richard Inwards, *Weather Lore: A Collection of Proverbs, Sayings, and Rules Covering the Weather*, 4th edn (London, 1950), p. 124.

8 Pao K. Wang, *Physics and Dynamics of Clouds and Precipitation* (Cambridge, 2013), p. ix.

9 Cited in Peter Moore, *The Weather Experiment: The Pioneers who Sought to See the Future* (London, 2015), p. 122.

10 Oliver Goldsmith, *A History of the Earth, and Animated Nature*, new edn (London, 1816), vol. I, p. 312.

11 Cited in Richard Hamblyn, *The Invention of Clouds: How an Amateur Meteorologist Forged the Language of the Skies* (London, 2001), p. 45.

12 Cited ibid., p. 62.

13 Luke Howard, *On the Modifications of Clouds* (London, 1804), p. 4.

14 Ibid., p. 8.

15 J. W. Goethe, *Italian Journey, 1786–1788*, trans. W. H. Auden and Elizabeth Mayer (Harmondsworth, 1970), p. 23.

16 Cited in Kurt Badt, *John Constable's Clouds*, trans. Stanley Godman (London, 1950), p. 18.

17 In D.F.S. Scott, *Some English Correspondents of Goethe* (London, 1949), pp. 51–4.

18 Ibid., pp. 48–50.

19 Cited in Hamblyn, *The Invention of Clouds*, pp. 219–20.

20 Cited in Badt, *John Constable's Clouds*, p. 15.

21 Ibid., p. 20.

22 Carl Gustav Carus, *Nine Letters on Landscape Painting: Written in the Years 1815–1824*, trans. David Britt (Los Angeles, CA, 2002), p. 114.

23 Ibid., p. 145.

24 Marie Lødrup Bang, *Johan Christian Dahl, 1788–1857: Life and Works* (Oslo, 1987), I, p. 79.

25 'The Cloud', in *Percy Bysshe Shelley: The Major Works*, ed. Zachary Leader (Oxford, 2003), pp. 461–3. See also J. E. Thornes, 'Luke Howard's Influence on Art and Literature in the Early Nineteenth Century', *Weather*, XXXIX (1984), p. 254. The poem was adapted as a silent movie in 1919, *The Cloud*, directed by W. A. van Scoy.

26 *Mégha Dúta; or Cloud Messenger; A Poem in the Sanskrit Language*, trans. Horace Hayman Wilson (Calcutta, 1813), p. 32.

27 Ralph Abercromby, 'Suggestions for an International Nomenclature of Clouds', *Quarterly Journal of the Royal Meteorological Society*, XIII (1887), p. 155.

28 Jules Verne, *Five Weeks in a Balloon*, trans. Arthur Chambers (London, 1926), p. 53.

29 Ralph Abercromby, *Seas and Skies in Many Latitudes; or, Wanderings in Search of Weather* (London, 1888), p. 218.
30 Ley, 'Clouds and Weather Signs', p. 107.
31 International Meteorological Committee, *Atlas International des Nuages/International Cloud Atlas/Internationaler Wolken-Atlas* (Paris, 1896), p. 13.
32 Albin J. Pollock, *The Underworld Speaks: An Insight to Vice-Crime-Corruption* (San Francisco, CA, 1935), p. 16. The subsequent entry reads: 'Cloud blowers: opium smokers; dopefiends; hopheads.' As was seen in Chapter One, the penultimate cloud of Buddhist enlightenment is also a contender for the original 'cloud nine'.
33 Brian Clegg, *Inflight Science: A Guide to the World from Your Airplane Window* (London, 2011), p. 95.
34 Air Ministry/Meteorological Office, *Cloud Atlas For Aviators*, 3rd edn (London, 1943), p. 2.
35 Ibid., pp. 8–10.
36 Anthony Le Grand, *An Entire Body of Philosophy, According to the Principles of the Famous Renate Des Cartes* (London, 1694), p. 215.
37 Goethe, *Italian Journey*, p. 32.
38 Philip Larkin, 'Cut Grass', in *High Windows* (London, 1974), p. 41.
39 Cited in Inwards, *Weather Lore*, p. 144.
40 Christopher C. Burt, *Extreme Weather: A Guide and Record Book* (New York, 2004), p. 162.
41 Luke Howard, letter to William Dillworth Howard, 17 July 1851, Newham Archives and Local Studies Library, London.

3 The Language and Literature of Clouds

1 Nicholson Baker, *Travelling Sprinkler* (London, 2013), pp. 69–70.
2 Cited in Jan Golinski, '"Exquisite Atmography": Theories of the World and Experiences of the Weather in a Diary of 1703', *British Journal for the History of Science*, XXXIV (2001), p. 168. Golinski has tentatively identified the author of the diary as Thomas Appletree of Worcestershire, an Oxford-educated lawyer and landowner.
3 Cited ibid., p. 156.
4 Hayman Rooke, *A Continuation of the Annual Meteorological Register, kept at Mansfield Woodhouse &c.* (Nottingham, 1802), p. 22.
5 William Clement Ley, 'Clouds and Weather Signs', in *Modern Meteorology: A Series of Six Lectures* (London, 1879), p. 104.
6 All from Robert MacFarlane, *Landmarks* (London, 2015), pp. 227–30.
7 The source is given as Thomas Hearne, *Remarks & Collections* (Oxford, 1885), vol. I, p. 112.

8 Cited in Thomas Sprat, *The History of the Royal-Society of London, For the Improving of Natural Knowledge* (London, 1667), p. 174.

9 Ibid., p. 177.

10 Ibid., pp. 177–8.

11 See J. A. Kington, 'The Societas Meteorologica Palatina: An Eighteenth-century Meteorological Society', *Weather*, XXIX (1974), pp. 416–26.

12 Cited in K. Khrgian, *Meteorology: A Historical Survey*, trans. Ron Hardin, 2nd edn (Jerusalem, 1970), p. 91.

13 F. H. Ludlam, 'History of Cloud Classification', in *Clouds of the World: A Complete Colour Encyclopedia*, ed. Richard Scorer (Newton Abbott, 1972), p. 17.

14 William Scoresby Jr., *An Account of the Arctic Regions* (Edinburgh, 1820), p. 250.

15 William Cobbett, *Rural Rides*, ed. George Woodcock (Harmondsworth, 1967), p. 123.

16 Robert FitzRoy, *Barometer and Weather Guide*, 2nd edn (London, 1859), p. 14.

17 Desmond King-Hele, *Shelley: His Thought and Work*, 2nd edn (London, 1971), p. 219.

18 *The Annual Review, and History of Literature, for 1804* (London, 1805), p. 900.

19 John Bostock, 'Remarks upon Meteorology', *Journal of Natural Philosophy, Chemistry and the Arts*, 26 (1810), pp. 2–9.

20 Luke Howard, 'Observations on Dr Bostock's Remarks upon Meteorology', *Journal of Natural Philosophy, Chemistry and the Arts*, 26 (1810), pp. 213–14.

21 Thomas Forster, 'The New Nomenclature of Clouds Explained', *Gentleman's Magazine*, LXXXI/2 (1811), p. 113.

22 Thomas Forster, 'Specimen of a New Nomenclature for Meteorological Science', *Gentleman's Magazine*, 86 (1816), p. 131.

23 Ibid., p. 132. Forster translated a number of other meteorological terms into English vernacular, such as 'Moon-burr' (corona) and 'Pole-streamer' (aurora).

24 Thomas Forster, *The Perennial Calendar, and Companion to the Almanack*, 3rd edn (London, 1823), pp. 93–4.

25 Luke Howard, *The Climate of London, deduced from Meteorological Observations* (London, 1818), vol. I, p. xxxii.

26 L. W. Gilbert, 'Versuch einer Naturgeschichte und Physik der Wolken', *Annalen der Physik*, 51 (1815), pp. 1–48.

27 Cited in Ralph Abercromby, 'Suggestions for an International Nomenclature of Clouds', *Quarterly Journal of the Royal Meteorological Society*, XIII (1887), p. 163.

28 Ralph Abercromby, 'On the Identity of Cloud Forms All Over the World', *Quarterly Journal of the Royal Meteorological Society*, XIII (1887), p. 141.

29 Andrés Poey, 'Sur deux nouveaux Types de Nuages observés à la Havane', *l'Annuaire de la Sociéte Météorologique de France*, XI (1863), p. 1.

30 William Clement Ley, *Cloudland: A Study on the Structure and Characters of Clouds* (London, 1894), pp. 26–7. The title was derived from a line in Coleridge's sonnet 'Fancy in Nubibus, or, The Poet in the Clouds' (1817).

31 Gustave Flaubert, *Bouvard and Pécuchet*, trans. A. J. Krailsheimer (Harmondsworth, 1976), p. 48.

32 Alan Clark, *Diaries* (London, 1993), p. 315.

33 In Alexander Pope, *The Major Works*, ed. Pat Rogers (Oxford, 2006), pp. 195–238. Pope's recipe bears some similarity to a passage by Joseph Addison in *The Spectator*, 592 (10 September 1714), in which he describes a series of weather effects ('a new Sett of Meteors') unveiled at Drury Lane Theatre: 'their Lightnings are made to flash more briskly than heretofore; their Clouds are also better furbelow'd, and more voluminous; not to mention a violent Storm locked up in a great Chest that is designed for the *Tempest*.'

34 'Mutability', in *Percy Bysshe Shelley: The Major Works*, ed. Zachary Leader (Oxford, 2003), p. 112. Shelley remarked in a letter to Thomas Love Peacock that 'I take great delight in watching the changes of the atmosphere.'

35 *The Letters of Percy Bysshe Shelley*, ed. Frederick L. Jones (Oxford, 1964), II, p. 25.

36 'The House of Clouds', in *The Works of Elizabeth Barrett Browning*, ed. Sandra Donaldson et al. (London, 2010), II, pp. 260–63.

37 Ibid., pp. 257–8.

38 Alexandra Harris, *Weatherland: Writers and Artists Under English Skies* (London, 2015), p. 234.

39 Baker, *Travelling Sprinkler*, pp. 69–70.

40 Lavinia Greenlaw, *A World Where News Travelled Slowly* (London, 1997), pp. 9–10.

41 'The Namer of Clouds', BBC Radio 4, 28 July 2013.

42 Carol Ann Duffy, 'Luke Howard, Namer of Clouds', in *The Bees* (London, 2011), p. 46. 'The Classification of Clouds', from Lesley Saunders's collection *Cloud Camera* (Reading, 2012), invokes a similar sentiment, declaring that:

> . . . Clouds
> defy the line, they will not be penned or pencilled,
> they fly and free-associate above my landscape

43 Billy Collins, 'Student of Clouds', in *Taking Off Emily Dickinson's Clothes: Selected Poems* (London, 2000), p. 31; Lewis Grassic Gibbon, *Cloud Howe*, ed. Tom Crawford (Edinburgh, 1989), p. xii.
44 Gibbon, *Cloud Howe*, p. 212.
45 The World Meteorological Organization Media Centre, via: http://public.wmo.int/en/media/news/international-cloud-atlas (accessed 20 November 2016).

4 Clouds in Art, Photography and Music

1 Alfred Stieglitz, 'How I Came to Photograph Clouds', *Amateur Photographer & Photography*, 56 (1923), p. 255.
2 William Gilpin, *Observations on the Western Parts of England, Relative Chiefly to Picturesque Beauty* (London, 1798), pp. 129–30.
3 Jonathan Swift, *A Tale of a Tub and Other Works*, ed. Angus Ross and David Woolley (Oxford, 1986), p. 16. How consciously was Swift echoing Mark Antony's complaint to Eros, from scene fourteen in Act IV of Shakespeare's *Antony and Cleopatra*:

> Sometimes we see a cloud that's dragonish;
> A vapour sometime like a bear or lion,
> A tower'd citadel, a pendent rock,
> A forkèd mountain, or blue promontory
> With trees upon't, that nod unto the world,
> And mock our eyes with air.

4 Mary Jacobus, 'Cloud Studies: The Visible Invisible', *Gramma: Journal of Theory and Criticism*, XIV (2006), p. 222.
5 Hubert Damisch, *A Theory of /Cloud/: Toward a History of Painting*, trans. Janet Lloyd (Stanford, CA, 2002), p. 174.
6 Leonardo da Vinci, *A Treatise on Painting*, trans. J. F. Rigaud (London, 1877), p. 144.
7 John E. Thornes, *John Constable's Skies: A Fusion of Art and Science* (Birmingham, 1999), p. 161.
8 John Ruskin, *Modern Painters*, I (London, 1843), cited ibid., p. 193.
9 Cited in Anthony Bailey, *John Constable: A Kingdom of his Own* (London, 2006), p. 126.
10 From C. R. Leslie, *Memoirs of the Life of John Constable* (1845), cited in Thornes, *John Constable's Skies*, p. 173. The quotation is from George Crabbe, 'The Lover's Journey' (1812); the italics are Leslie's own.
11 Cited in Thornes, *John Constable's Skies*, p. 51.
12 Ibid., p. 90.
13 Kurt Badt, *John Constable's Clouds*, trans. Stanley Godman (London, 1950), p. 48.

14 John Ruskin, 'Preface', *Modern Painters*, I, 2nd edn (London, 1844), p. xxxvii.

15 Ruskin, *Modern Painters*, V (London, 1860), p. 144.

16 Ibid., p. 112.

17 In Alexandra Harris, *Weatherland: Writers and Artists Under English Skies* (London, 2015), pp. 315–16.

18 Ruskin, *Modern Painters*, V, pp. 127–8.

19 John Ruskin, *The Storm-cloud of the Nineteenth Century* (London, 1884), pp. 58–9.

20 Ibid., p. 32.

21 Virginia Woolf, *Orlando: A Biography* [1928], ed. Brenda Lyons (London, 1993), pp. 157–8.

22 See 'Cloud', in Christoph Grunenberg and Darren Phi, eds, *Magritte A–Z* (London, 2011), pp. 32–5.

23 William Clement Ley, 'Clouds and Weather Signs', in *Modern Meteorology: A Series of Six Lectures* (London, 1879), p. 104.

24 Alfred Angot, 'Amateur Cloud Photography', *Nature*, LIII (1896), p. 230.

25 Ibid., p. 232.

26 Norman Lockyer, 'The Photographic Observation of Clouds', *Nature*, LV (1897), pp. 322–3.

27 Ibid., p. 324.

28 G. M. Whipple, 'A Brief Notice Respecting Photography in Relation to Meteorological Work', *Quarterly Journal of the Royal Meteorological Society*, XVI (1890), p. 145. See also Jennifer Tucker, *Nature Exposed: Photography as Eyewitness in Victorian Science* (Baltimore, MD, 2005), pp. 145–9.

29 Lockyer, 'The Photographic Observation of Clouds', p. 324.

30 Ralph Abercromby, 'Suggestions for an International Nomenclature of Clouds', *Quarterly Journal of the Royal Meteorological Society*, XIII (1887), pp. 154–5.

31 Birt Acres, 'Some Hints on Photographing Clouds', *Quarterly Journal of the Royal Meteorological Society*, XXI (1895), pp. 160–61.

32 Ibid., pp. 154–5.

33 Ralph Abercromby, 'On the Identity of Cloud Forms all over the World', *Quarterly Journal of the Royal Meteorological Society*, XIII (1887), p. 146.

34 H. H. Hildebrandsson, W. Köppen and G. Neumayer, *Wolken-Atlas – Atlas des Nuages – Cloud-Atlas – Moln-Atlas* (Hamburg, 1890), p. 3.

35 Ibid.

36 Stieglitz, 'How I Came to Photograph Clouds', p. 255.

37 H. H. Hildebrandsson et al., *Wolken-Atlas*, p. 3.

38 Angot, 'Amateur Cloud Photography', p. 232.

39 Gavin Pretor-Pinney, *Hot Pink Flying Saucers, and Other Clouds from the Cloud Appreciation Society* (New York, 2007), p. 3.

40 *The Complete Peanuts* (Seattle, WA, 2006), V, p. 254.

41 David Hume, *Dialogues and Natural History of Religion*, ed. J.C.A. Gaskin (Oxford, 1993), p. 141.

42 'Fancy in Nubibus, or, The Poet in the Clouds', in *The Poems of Samuel Taylor Coleridge*, ed. Ernest Hartley Coleridge (Oxford, 1949), p. 435.

43 Cited in Richard Holmes, *Coleridge: Early Visions* (London, 1989), p. 278.

44 Aristophanes, *The Clouds*, in *Lysistrata and Other Plays*, trans. Alan H. Sommerstein (Harmondsworth, 1973), p. 127. Bill Watterson, *Calvin and Hobbes*, 16 June 1993. See Jamie Heit, *Imagination and Meaning in Calvin and Hobbes* (Jefferson, NC, 2012), p. 45.

45 Gavin Pretor-Pinney, *Clouds That Look Like Things: From the Cloud Appreciation Society* (London, 2012), p. 8. Today, smartphones have made everyone a photographer, but it's also the case that, as the technology changes, so do the associated challenges: autofocus systems inside cameras and phones often fail to find a target area when pointed at the sky, producing disappointingly flat or out-of-focus cloudscapes, while overexposure remains the Achilles heel of cloud photography. As many of the images to be found on the Cloud Appreciation Society's website confirm, effective cloud photography requires an element of luck and patience, but it also depends upon a high degree of technical skill.

46 Cited in William Wellman Jr., *The Man and His Wings: William A. Wellman and the Making of the First Best Picture* (Westport, CT, 2006), p. 118.

47 John Park Finley, *Tornadoes: What they are and How to Observe them, with Practical Suggestions for the Protection of Life and Property* (New York, 1887), p. 7.

48 Ibid., p. 33.

49 Cited in Sean Potter, 'April 26, 1884: Earliest Known Tornado Photograph', *Weatherwise*, LXIII/2 (2010), p. 13.

50 Finley, *Tornadoes*, p. 36.

51 Quoted in Aljean Harmetz, *The Making of the Wizard of Oz* (New York, 1998), p. 247.

52 Ibid.

53 Paul N. Edwards, *A Vast Machine: Computer Models, Climate Data, and the Politics of Global Warming* (Cambridge, MA, 2010), p. 219.

54 Ken Croswell, 'Water Clouds Tentatively Detected Just 7 Light-years from Earth', *Science*, 25 August 2014, via http://news.sciencemag.org (accessed 20 September 2015).

55 James Glaisher et al., *Travels in the Air* (London, 1871), pp. 173–4.

56 Ibid., p. 44.
57 Mark Strand, '14', in Wendy Mark and Mark Strand, *89 Clouds* (New York, 1999).
58 In Hugh Cornwell and Jim Drury, *The Stranglers: Song by Song* (London, 2001), p. 96. The track was also recorded in a Swedish-language version, entitled 'Sverige'.
59 Cited in Ron Moy, *Kate Bush and 'Hounds of Love'* (Aldershot, 2007), p. 46.
60 Quoted in Myron Sharaf, *Fury on Earth: A Biography of Wilhelm Reich* (London, 1983), p. 379.
61 Cited in Simon Trezise, ed., *The Cambridge Companion to Debussy* (Cambridge, 2003), p. 104.
62 Ibid., p. 105.
63 Karl Popper, *Of Clouds and Clocks: An Approach to the Problem of Rationality and the Freedom of Man* (St Louis, MI, 1966), p. 2.
64 Ibid., p. 4.
65 Richard Steinitz, *György Ligeti: Music of the Imagination* (London, 2003), p. 199.
66 Ibid., pp. 199–200.

5 Future Clouds

1 Christopher Barnatt, *A Brief Guide to Cloud Computing* (London, 2010), p. 3.
2 Ibid., p. 216.
3 Cited in Walter Isaacson, *Steve Jobs* (New York, 2011), p. 533.
4 Cited in Johnny Magdaleno, '"The Cloud" is Actually a Tangible Thing – And This is What it Looks Like', via http://thecreatorsproject.vice.com/blog, 22 May 2014.
5 Cited in James Rodger Fleming, *Fixing the Sky: The Checkered History of Weather and Climate Control* (New York, 2010), p. 62. Particulates in rising gunsmoke may well act as 'nucleation sites' for cloud droplets, leading to the phenomenon of post-battle rain.
6 Laurie Lee, *Cider With Rosie* (London, 1959), p. 37.
7 Gavin Pretor-Pinney, *The Cloudspotter's Guide* (London, 2006), p. 259.
8 Cited in Cynthia Barnett, *Rain: A Natural and Cultural History* (New York, 2015), p. 177.
9 Cited in Fleming, *Fixing the Sky*, p. 180.
10 Owen Gibson, 'Olympic Opening Ceremony will Recreate Countryside with Real Animals', *The Guardian*, 12 June 2012, p. 7.
11 Cited in Laurence Boisson de Chazournes, *Fresh Water in International Law* (Oxford, 2013), p. 46.

12 'China Rain-making Creates a Storm', BBC News, 14 July 2004, via http://news.bbc.co.uk (accessed 24 September 2015). The Chinese government claims that some 560,000 cloud-seeding missions over the past ten years have generated nearly 500 billion tonnes of rain.

13 De Chazournes, *Fresh Water in International Law*, p. 48.

14 David J. Travis et al., 'Contrails Reduce Daily Temperature Range', *Nature*, 418 (2002), p. 601.

15 Patrick Minnis et al., 'Contrails, Cirrus Trends, and Climate', *Journal of Climate*, XVII/8 (2004), pp. 1671–85.

16 E. A. Irvine et al., 'A simple framework for assessing the trade-off between the climate impact of aviation carbon dioxide emissions and contrails for a single flight', *Environmental Research Letters*, IX/6 (2014), via http://iopscience.iop.org (accessed 22 September 2015).

17 Andrew J. Heymsfield et al., 'Formation and Spread of Aircraft-induced Holes in Clouds', *Science*, 333 (2011), pp. 77–81.

18 William Clement Ley, 'Clouds and Weather Signs', in *Modern Meteorology: A Series of Six Lectures* (London, 1879), p. 103.

19 Timothy Donnelly, *The Cloud Corporation* (Seattle, WA, 2010), p. 31.

20 Cited in Elizabeth Diller and Ricardo Scofidio, *Blur: The Making of Nothing* (New York, 2002), p. 172.

21 Martin Caiger-Smith, *Antony Gormley* (London, 2010), p. 108.

22 Cited in Diller and Scofidio, *Blur*, p. 192.

23 Cited in Caiger-Smith, *Antony Gormley*, p. 102.

24 In *The Poetical Works of William Wordsworth*, ed. Ernest de Selincourt and Helen Darbishire (Oxford, 1946), III, p. 84.

25 Quoted in Christine Ottery, 'Deck the Halls with Clouds', *New Scientist*, 214 (9 June 2012), p. 49.

26 Jeffrey L. Kosky, *Arts of Wonder* (Chicago, IL, 2013), p. 113. In 1969 Turrell created sky drawings using coloured skywriting smoke and cloud-seeding materials.

27 Craig Adcock, *James Turrell: The Art of Light and Space* (Berkeley, CA, 1990), p. 171.

28 Ibid., p. 207.

29 *Climate Change 2007: The Physical Science Basis. Contribution of Working Group I to the Fourth Assessment Report of the Intergovernmental Panel on Climate Change* (Cambridge, 2007), p. 116.

30 'Clouds and Aerosols', in *Climate Change 2013: The Physical Science Basis. Contribution of Working Group I to the Fifth Assessment Report of the Intergovernmental Panel on Climate Change* (Cambridge, 2013), p. 573.

31 Amy C. Clement et al., 'Observational and Model Evidence for Positive Low-Level Cloud Feedback', *Science*, 325 (2009), pp. 460–64.

32 David M. Romps et al., 'Projected Increase in Lightning Strikes in the United States Due to Global Warming', *Science*, 346 (2014), pp. 851–4.

33 William Bryant Logan, *Air: The Restless Shaper of the World* (New York, 2012), pp. 64–5.

34 Ralph Abercromby, 'Suggestions for an International Nomenclature of Clouds', *Quarterly Journal of the Royal Meteorological Society*, XIII (1887), p. 163.

SELECT BIBLIOGRAPHY

Audeguy, Stéphane, *The Theory of Clouds*, trans. Timothy Bent
   (New York, 2007)
Badt, Kurt, *John Constable's Clouds*, trans. Stanley Godman
   (London, 1950)
Barnett, Cynthia, *Rain: A Natural and Cultural History* (New York, 2015)
Boia, Lucian, *The Weather in the Imagination*, trans. Roger Leverdier
   (London, 2005)
Broglio, Ron, *Technologies of the Picturesque: British Art, Poetry, and
   Instruments, 1750–1830* (Lewisburg, PA, 2008)
Clegg, Brian, *Inflight Science: A Guide to the World from Your Airplane
   Window* (London, 2011)
—, *Exploring the Weather* (London, 2013)
Connor, Steven, *The Matter of Air: Science and the Art of the Ethereal*
   (London, 2010)
Damisch, Hubert, *A Theory of /Cloud/: Toward a History of Painting*,
   trans. Janet Lloyd (Stanford, CA, 2002)
Day, John A., *The Book of Clouds* (New York, 2006)
Dunlop, Storm, *Photographing Weather* (London, 2007)
—, *Meteorology Manual: The Practical Guide to the Weather* (Yeovil, 2014)
Edwards, Paul N., *A Vast Machine: Computer Models, Climate Data, and
   the Politics of Global Warming* (Cambridge, MA, 2010)
Hamblyn, Richard, *The Invention of Clouds: How an Amateur
   Meteorologist Forged the Language of the Skies* (London, 2001)
—, 'A Celestial Journey', *TateEtc*, 5 (2005)
—, *The Cloud Book* (Newton Abbott, 2008)
—, *Extraordinary Clouds* (Newton Abbot, 2009)
Harris, Alexandra, *Weatherland: Writers and Artists Under English Skies*
   (London, 2015)
Hawes, Louis, 'Constable's Sky Sketches', *Journal of the Warburg and
   Courtauld Institute*, 32 (1969), pp. 344–65

Hill, Jonathan, *Weather Architecture* (Abingdon, 2012)

Howard, Luke, *Seven Lectures on Meteorology* (Pontefract, 1837)

—, *On the Modifications of Clouds*, 3rd edn (London, 1865)

Inwards, Richard, *Weather Lore: A Collection of Proverbs, Sayings, and Rules Covering the Weather*, 4th edn (London, 1950)

Jacobus, Mary, 'Cloud Studies: The Visible Invisible', *Gramma: Journal of Theory and Criticism*, XIV (2006), pp. 219–47

Janković, Vladimir, *Reading the Skies: A Cultural History of English Weather, 1650–1820* (Manchester, 2001)

Jha, Alok, *The Water Book* (London, 2015)

Ley, William Clement, *Cloudland: A Study on the Structure and Characters of Clouds* (London, 1894)

Logan, William Bryant, *Air: The Restless Shaper of the World* (New York, 2012)

Moore, Peter, *The Weather Experiment: The Pioneers who Sought to See the Future* (London, 2015)

Morris, Edward, ed., *Constable's Clouds* (Edinburgh, 2000)

Pretor-Pinney, Gavin, *The Cloudspotter's Guide* (London, 2006)

—, *Hot Pink Flying Saucers, and Other Clouds from the Cloud Appreciation Society* (New York, 2007)

—, *Clouds That Look Like Things: From the Cloud Appreciation Society* (London, 2012)

Ruskin, John, *The Storm-cloud of the Nineteenth Century* (London, 1884)

Scorer, Richard, *Clouds of the World: A Complete Colour Encyclopedia* (Newton Abbott, 1972)

Stephens, Graeme L., 'The Useful Pursuit of Shadows', *American Scientist*, 91 (2003), pp. 442–9

Thornes, John E., *John Constable's Skies: A Fusion of Art and Science* (Birmingham, 1999)

Völter, Helmut, ed., *Wolkenstudien | Cloud Studies | Études des Nuages* (Leipzig, 2011)

ASSOCIATIONS AND WEBSITES

The Cloud Appreciation Society
http://cloudappreciationsociety.org

Cloudman™ (website of the late Dr John Day)
www.cloudman.com

Clouds Online / Wolken Online
www.clouds-online.com

For Spacious Skies
www.forspaciousskies.com

The International Cloud Atlas
http://public.wmo.int/en/media/news/international-cloud-atlas

Met Office Clouds page
www.metoffice.gov.uk/learning/clouds

National Weather Service Online Weather School
www.srh.noaa.gov/jetstream/clouds/clouds_intro.htm

The Royal Meteorological Society
www.rmets.org

The Tornado and Storm Research Organisation
www.torro.org.uk

ACKNOWLEDGEMENTS

My thanks are due to Daniel Allen and Michael Leaman for commissioning and editing this book, and to Reaktion's in-house editors and designers for their patient work on the manuscript. It's a pleasure to thank the staff of the British Library, the National Meteorological Library and Archive, the Science Museum Library and the Senate House Library, University of London, where much of this book was researched. I am also very grateful to the friends and colleagues with whom I have discussed clouds and weather over the years, including Jon Adams, Julia Bell, Martin John Callanan, Gregory Dart, Markman Ellis, Alexandra Harris, Vladimir Janković, Esther Leslie, Mark Maslin, Peter Moore, Penny Newell, Michael Newton, Gavin Pretor-Pinney, Sam van Schaik, Emily Senior, Liane Strauss, Colin Teevan, John Thornes and Sue Wiseman, as well as to Nina Whiteman, who kindly allowed me to reproduce a sample of her score for *The Modifications of Clouds*.

For the opportunity to discuss some of the material while it was still being written, I would like to thank my fellow speakers and audience members at 'Clouds: Objects, Metaphor, Phenomena', Birkbeck Arts Week, May 2014; 'Escape to the Clouds', the Cloud Appreciation Society's hugely enjoyable tenth anniversary conference in September 2015; and at 'Art and Skies', a one-day conference co-organized by the Royal Meteorological Society and Tate Britain in December 2015. My warmest thanks and love, however, are for Jo, Ben and Jessie Hamblyn, to whom this book is fondly dedicated.

PHOTO ACKNOWLEDGEMENTS

The author and publishers wish to express their thanks to the below sources of illustrative material and/or permission to reproduce it. Locations of some artworks are also given below.

Ashmolean Museum, Oxford: p. 129; photos by or courtesy of the author: pp. 9, 65, 70, 71, 74, 81, 82, 113, 116, 120, 144, 149, 167, 170–71, 186, 193, 194–5 (foot), 199, 200, 204–5; from Thomas Baldwin, *Airopaidia: Containing the Narrative of a Balloon Excursion from Chester, the 8th of September, 1785 . . .* (Chester, 1786): p. 36; photo Beckachester: p. 187 (this file is licensed under the Creative Commons Attribution-Share Alike 4.0 International license: any reader is free to share – to copy, distribute and transmit the work, or to remix – to adapt the work, under the following conditions: you must attribute the work in the manner specified by the author or licensor (but not in any way that suggests that they endorse you or your use of the work); photos Julia Bell: pp. 20, 40, 119, 163; photo Bering Land Bridge National Preserve, Nome, Alaska: p. 99 (this file is licensed under the Creative Commons Attribution 2.0 Generic license: any reader is free to share – to copy, distribute and transmit the work, or to remix – to adapt the work, under the following conditions: you must attribute the work in the manner specified by the author or licensor (but not in any way that suggests that they endorse you or your use of the work); reproduced by kind permission of the artist (Martin John Callanan), photo © MJC 2009: p. 158 (top); photo Kevin Cho [Kee Pil Cho]: p. 86 (this file is licensed under the Creative Commons Attribution-Share Alike 3.0 Unported license: any reader is free to share – to copy, distribute and transmit the work, or to remix – to adapt the work, under the following conditions: you must attribute the work in the manner specified by the author or licensor (but not in any way that suggests that they endorse you or your use of the work); photo cking: pp. 42–3 (This file is licensed under the Creative Commons Attribution 2.0 Generic license: any reader is free to

share – to copy, distribute and transmit the work, or to remix – to adapt
the work, under the following conditions: you must give appropriate credit,
provide a link to the license, and indicate if changes were made – you may
do so in any reasonable manner, but not in any way that suggests the
licensor endorses you or your use – with no additional restrictions — you
may not apply legal terms or technological measures that legally restrict
others from doing anything the license permits); Cleveland Museum of
Art, Ohio/Mr and Mrs. William H. Marlatt Fund /Bridgeman Images:
p. 23; from Alexander Cozens, *A New Method of Assisting the Invention
in Drawing Original Compositions of Landscape* (London, n.d. [1785/86]):
p. 128; Davis Museum and Cultural Center, Wellesley College, Massa-
chusetts (gift of Mrs Leeds A. Wheeler – Marion Eddy, Class of 1924)/
Bridgeman Images: pp. 32–3; photo Denni: p. 73 (this file is licensed under
the Creative Commons Attribution-Share Alike 2.5 Generic license: any
reader is free to share – to copy, distribute and transmit the work, or to
remix – to adapt the work, under the following conditions: you must attribute the work in the manner specified by the author or licensor (but not
in any way that suggests that they endorse you or your use of the work);
from Henri Dore, 'Vie illustrée du Bouddha Çakyamouni', vol. xv of his
Recherches sur les superstitions en Chine (Chang-Hai, 1929): p. 14; photo
Earth Sciences and Image Analysis Laboratory, Johnson Space Center/
NASA: pp. 190–91; photo EllsworthC: p. 192; (this file is licensed under the
Creative Commons Attribution 3.0 Unported license: any reader is free
to share – to copy, distribute and transmit the work, or to remix – to adapt
the work, under the following conditions: you must attribute the work in
the manner specified by the author or licensor (but not in any way that
suggests that they endorse you or your use of the work); photo Emilio
Segre Visual Archives/American Institute of Physics/Science Photo
Library: p. 176; photo Simon A. Eugster: p. 84 (this file is licensed under
the Creative Commons Attribution-Share Alike 3.0 Unported license: any
reader is free to share – to copy, distribute and transmit the work, or to
remix – to adapt the work, under the following conditions: you must attribute the work in the manner specified by the author or licensor (but not
in any way that suggests that they endorse you or your use of the work); from
John Park Finley, *Tornadoes: What They Are and How to Observe Them:
With Practical Suggestions for the Protection of Life and Property* (New York,
1887), courtesy History of Science Collections, University of Oklahoma
Libraries: pp. 150, 151, 153; from Camille Flammarion, *L'Atmosphere,
description des grands phénomènes de la nature* (London, 1873): p. 30; from
Thomas Forster, *Researches about Atmospheric Phaenomena . . .* (London,
1815): p. 102; from James Glaisher, ed., *Travels in the Air* (London, 1871):
p. 38; photo Grant W. Goodge: p. 27; from H. A. Guerber, *Myths of the
Norsemen: From the Eddas and Sagas* (London, 1909): p. 18; Hamburger
Kunsthalle: pp. 57, 61; from J. G. Heck, *Iconographic Encyclopædia of Science,*

Literature, and Art (New York, 1851): pp. 96, 97; photo Carol M. Highsmith/ Library of Congress, Washington, DC (Prints and Photographs Division): p. 77; from Luke Howard, 'On the Modifications of Clouds, and on the Principles of their Production, Suspension, and Destruction, being the Substance of an Essay read before the Askesian Society in the Session 1802–3', *Philosophical Magazine*, XVI/62 (1803): p. 50; from The International Meteorological Committee, *International Cloud-Atlas, published by order of the Committee by H. Hildebrandsson, A. Riggenbach, L. Teisserenc de Bort, members of the Clouds Commission* (Paris, 1896): p. 140; photo King›s College, London/Science Photo Library: p. 69; photo Ralph F. Kresge/ National Oceanic and Atmospheric Administration: p. 72; Kunsthistorisches Museum, Vienna: p. 19; from [The Meteorological Society], *Modern Meteorology: A Series of Six Lectures . . .* (London, 1879): p. 108; Library of Congress, Washington, DC: p. 91; photos Library of Congress, Washington, DC (Prints and Photographs Division): pp. 10, 41, 91; photo Livioandronico2013: p. 123 (this file is licensed under the Creative Commons Attribution-Share Alike 4/0 International license: any reader is free to share – to copy, distribute and transmit the work, or to remix – to adapt the work, under the following conditions: you must attribute the work in the manner specified by the author or licensor (but not in any way that suggests that they endorse you or your use of the work); from Elias Loomis, *A Treatise on Meteorology: with a Collection of Meteorological Tables* (New York, 1868): p. 69; courtesy Jo Lynch: p. 146; from Elié Margollé and Frédéric Zurcher, *Les Météores* (Paris, 1869): p. 90; photos Mary Evans Picture Library: pp. 11, 14; photo Mcostar: p. 189 (this file is licensed under the Creative Commons Attribution-Share Alike 3.0 Unported license: any reader is free to share – to copy, distribute and transmit the work, or to remix – to adapt the work, under the following conditions: you must attribute the work in the manner specified by the author or licensor (but not in any way that suggests that they endorse you or your use of the work); photo Ave Maria Mõistlik: p. 115 (this file is licensed under the Creative Commons Attribution-Share Alike 3.0 Unported license: any reader is free to share – to copy, distribute and transmit the work, or to remix – to adapt the work, under the following conditions: you must attribute the work in the manner specified by the author or licensor (but not in any way that suggests that they endorse you or your use of the work); Metropolitan Museum of Art: pp. 8, 124, 126, 136; Musée du Louvre, Paris: p. 125; Museum der bildenden Künste, Leipzig: p. 56; photos National Aeronautics & Space Administration: pp. 75, 78, 161; photo National Aeronautics & Space Administration/Earth Sciences and Image Analysis Laboratory at Johnson Space Centre: pp. 184–5; photo © National Aeronautics & Space Administration/Science Photo Library – all rights reserved: pp. 158 (foot), 159, 160 (top), 195 (top); National Museum of India, New Delhi/Bridgeman Images: p. 17; National Oceanic and Atmospheric Administration Photo

Library: pp. 90, 152, 155; National Oceanic and Atmospheric Administration Photo Library (National Weather Service Collection): pp. 27, 157; photo National Oceanic and Atmospheric Administration Photo Library, NOAA Central Library – OAR/ERL/National Severe Storms Laboratory (NSSL): p. 80; photo paleo_bear: p. 31 (this file is licensed under the Creative Commons Attribution 2.0 Generic license: any reader is free to share – to copy and redistribute the material in any medium or format, or to adapt – to remix, transform, and build upon the material, for any purpose, even commercially, under the following conditions: — you must give appropriate credit, provide a link to the license, and indicate if changes were made - you may do so in any reasonable manner, but not in any way that suggests the licensor endorses you or your use; and you may not apply legal terms or technological measures that legally restrict others from doing anything the license permits; photo David Pollack/Corbis Historical via Getty Images: p. 39; photo Lamont Poole/National Aeronautics & Space Administration: p. 85; private collection (photo Christie's Images © Bridgeman Images, © ADAGP, Paris and DACS, London, 2016): p. 134; from *Puck*, vol. LXX/1810 (8 November 1911): p. 10; photo robertharding/ Alamy: p. 16; from Hayman Rooke, *A Continuation of the Annual Meteorological Register, Kept at Mansfield Woodhouse, from the Year 1800 to the End of the Year 1801* (Nottingham, 1802): p. 88; Royal Meteorological Society (currently on loan to the Science Museum, London – photo © Science Museum/Science & Society Picture Library – all rights reserved): p. 52; from Nicola Sabbatini, *Pratica de Fabricar Scene, e Machine ne' Teatri* (Ravenna, 1638): pp. 168, 169; from *La Science illustrée*, vol. 11/347 (21 July 1894): p. 11; photos © Science Museum/Science & Society Picture Library – all rights reserved: pp. 50, 53, 55; photo Adrian Simpson: p. 79; photo Chris Spannagle/National Oceanic and Atmospheric Administration Photo Library, NOAA Central Library – OAR/ERL/National Severe Storms Laboratory (NSSL): p. 206; from Thomas Sprat, *The History of the Royal-Society of London, for the improving of natural knowledge* (London, 1667): p. 94; photo stokpic.com: p. 175; photo U.S. Air Force: p. 196; photos U.S. Department of Defence: pp. 194 (top), 208–9; photo U.S. Navy/Photographer's Mate 2nd Class Daniel J. McLain: p. 198; Victoria & Albert Museum, London: p. 127; photo by Visual China Group via Getty Images: p. 179; photo Nick Webb: pp. 182–3 (this file is licensed under the Creative Commons Attribution 2.0 Generic license: any reader is free to share – to copy, distribute and transmit the work, or to remix – to adapt the work, under the following conditions: you must attribute the work in the manner specified by the author or licensor (but not in any way that suggests that they endorse you or your use of the work); from Harold F. B. Wheeler, ed., *The Book of Knowledge: A Pictured Encyclopaedia for Readers of all Ages*, vol. 11 (London, [1926]): p. 107; by kind permission of Nina Whiteman: p. 172; photo Wojtow: p. 29 (this file is licensed under the Creative Commons Attribution-Share Alike 3.0 Unported license: any reader is free to share – to copy, distribute and transmit the work, or to remix – to adapt the work, under the following conditions: you must attribute the work in the manner specified by the author or licensor (but not in any way that suggests that they endorse you or your use of the work).

INDEX